What Makes a Good Farm for Wildlife?

Dedication

To the wonderful and highly dedicated volunteers from the Canberra Ornithologists Group (COG).

To the many farmers who have allowed us to study on their land – and provide us with practical insights that can never be learned in a university.

What Makes a Good Farm for Wildlife?

David Lindenmayer *(Lead Author)*, Sam Archer, Philip Barton, Suzi Bond, Mason Crane, Philip Gibbons, Geoff Kay, Christopher MacGregor, Adrian Manning, Damian Michael, Rebecca Montague-Drake, Nicola Munro, Rachel Muntz and Karen Stagoll

PUBLISHING

National Library of Australia Cataloguing-in-Publication entry

What makes a good farm for wildlife?/David Lindenmayer ... [et al.].

9780643100312 (pbk.)
9780643101623 (eBook)
9780643102217 (ePub)

Includes bibliographical references and index.
Biodiversity conservation – Australia.
Land use, Rural – Australia.
Agricultural conservation – Australia.
Wildlife conservation – Australia.
Plant diversity conservation – Australia.
Fragmented landscapes – Environmental aspects – Australia.
Environmental management – Australia.

Lindenmayer, David.

333.95160994

Published by

CSIRO PUBLISHING
36 Gardiner Road, Clayton VIC 3168
Private Bag 10, Clayton South VIC 3169
Australia

Telephone:	[+613] 9545 8555
Local call:	1300 788 000 (Australia only)
Fax:	+61 3 9662 7555
Email:	csiropublishing@csiro.au
Web site:	www.publishing.csiro.au

Front cover: Photo by David Lindenmayer

Back cover from top: Eastern Yellow Robin (Graeme Chapman); Squirrel Glider (Esther Beaton); Curl Snake (Damian Michael); Snake-eyed Skink (Damian Michael)

Set in Adobe Minion Pro 11/13.5 and Adobe Helvetica Neue LT
Edited by Janet Walker
Cover design by Samantha Duque
Text design by James Kelly
Typeset by Desktop Concepts Pty Ltd, Melbourne
Index by Indexicana
Printed by Ingram Lightning Source

CSIRO PUBLISHING publishes and distributes scientific, technical and health science books and journals from Australia to a worldwide audience and conducts these activities autonomously from the research activities of the Commonwealth Scientific and Industrial Research Organisation (CSIRO). The views expressed in this publication are those of the author(s) and do not necessarily represent those of, and should not be attributed to, the publisher or CSIRO.

Feb26_RP_ILS

Contents

Biographies of the authors

David Lindenmayer

David Lindenmayer is a Research Professor at the Fenner School of Environment and Society at The Australian National University where he has worked since 1992. He has been part of a research team in the temperate woodlands of southern New South Wales since 1997. He has special expertise in the biology and management of birds and arboreal marsupials.

Sam Archer

Sam Archer is a fifth generation livestock producer from Gundagai, New South Wales. In 2002 his farm was selected as a pilot site for the Australian Government's National Environmental Stewardship Program. In 2008 he was awarded a Nuffield Farming Scholarship and travelled overseas researching non-government funded environmental stewardship schemes. He is Chair of Murrumbidgee Landcare and a member of the Australian Farm Institute's research advisory committee.

Philip Barton

Philip Barton has undertaken studies in biology and ecology, and has a particular interest in insect ecology. His research for his PhD focuses on the spatial and functional ecology of beetles in eucalypt woodlands.

Suzi Bond

Suzi Bond has a strong interest in wildlife conservation and ecology. She is currently researching woodland birds in plantings and woodland remnants in agricultural landscapes for her PhD.

Mason Crane

Mason Crane started working on temperate woodland research with David Lindenmayer's team in 1999. Growing up in Gundagai, New South Wales, he developed a passion for biodiversity conservation in rural landscapes and an intimate understanding of rural communities. He has specialist expertise in woodland ecology and management, and in recent years has focused his research on possum and glider conservation in agricultural landscapes.

Philip Gibbons

Phil Gibbons gave up a burgeoning career in an abbatoir to study forestry and conservation biology. He has worked for 20 years across a range of organisations including the Victorian and New South Wales Governments and CSIRO, is a former board member of Greening Australia and has worked as a consultant for several government departments including the Chinese Government. He is currently an academic at The Australian National University.

Geoff Kay

Geoff Kay has worked in David Lindenmayer's research team at The Australian National University's Fenner School of Environment and Society since 2006. His background spans the fields of botany, ecology, herpetology, genetics and forestry, and he has a special interest in conserving woodland biodiversity in rural landscapes. He is currently working on a major woodland stewardship project across New South Wales and Queensland.

Christopher MacGregor

Christopher MacGregor is a Senior Research Officer at the Fenner School of Environment and Society at The Australian National University. He has been working as a wildlife ecologist with David Lindenmayer's research team since 1998. A keen ornithologist and botanist, he has experience in wildlife research on the South Coast and South West Slopes of New South Wales and the Central Highlands of Victoria.

Adrian Manning

Adrian Manning is a Fellow at the Fenner School of Environment and Society at The Australian National University. His research interests are in conservation biology, landscape ecology and restoration ecology. His work ranges from applied and field-based research in Australia, to research on general ecological issues, including climate adaptation, ecosystem restoration and conceptual theories in landscape ecology. Adrian manages a large research project investigating ways of improving Box-Gum grassy woodlands for biodiversity. As part of this, he is working with the ACT Government and colleagues to return locally extinct native mammals to Mulligans Flat Woodland Sanctuary in Canberra.

Damian Michael

Damian Michael is a Senior Research Officer at the Fenner School of Environment and Society at The Australian National University. He has been part of David Lindenmayer's research team since 2000, and has specialist expertise in botany, ecology and herpetology. In 2010 Damian completed a PhD on the ecology and conservation of reptiles in agricultural landscapes with a focus on managing rocky outcrops.

Rebecca Montague-Drake

Rebecca Montague-Drake is a Senior Research Officer at the Fenner School of Environment and Society at The Australian National University. She joined David Lindenmayer's research team in 2005, and is based in Gundagai, New South Wales. She is passionate about working to improve biodiversity conservation in woodland environments. Her particular areas of interest include small mammals and birds.

Nicola Munro

Nicola Munro completed her PhD in 2009 on the biodiversity value of revegetation in agricultural areas. She has a background in Australian arid zone ecology, mammal reintroductions and restoration ecology. She is particularly interested in small mammals, birds and plants, especially in the arid zone, the alps, and in the woodlands and forests of southern Australia. She is passionate about working with farmers and land managers to find a more sustainable future for all.

Rachel Muntz

Rachel Muntz has provided research and administrative assistance since 2007 to David Lindenmayer's research group at the Fenner School of Environment and Society at The Australian National University. She regularly contributes to scientific books and papers in an editorial capacity and has occasional co-authorship roles. Since joining the research team, she has become a keen birdwatcher.

Karen Stagoll

Karen Stagoll has a strong interest in wildlife conservation, and has a background in ecology, biology and environmental law. She is currently undertaking a PhD on the conservation and management of habitat for woodland birds in urban and peri-urban areas.

Preface

The conservation of biodiversity is now well recognised as a key part of ecologically sustainable natural resource management. This includes grazing and cropping enterprises throughout the agricultural heartland of south-eastern Australia – a set of regions which coincide with large parts of the nation's temperate woodlands. The conservation of these temperate woodlands on private properties is extremely important as they support many threatened species of mammals, birds and plants.

Temperate woodland landscapes have been modified extensively over the past 220 years, making the development of ways to integrate conservation with agriculture and livestock grazing a significant research and management challenge. In many ways, the conservation of biodiversity in these areas will be a major test of Australian's ability to develop farming practices that are both financially and ecologically sustainable.

For much of the past century, Australian temperate woodlands have received only limited attention from ecologists. This has changed in the past two decades. Over the last 15 years, a team of scientists from the Fenner School of Environment and Society at The Australian National University has been identifying ways to improve the conservation of wildlife on farms. The work has spanned studies of different vegetation types ranging from remnant native woodland to replanted native vegetation. It has also included places not traditionally associated with a wealth of biodiversity such as farm dams and rocky outcrops. Our research also has encompassed many different groups of organisms – from plants and insects through to reptiles, frogs, birds and mammals.

This new book is a compilation of findings from our past decade of research. It is about biodiversity on farms and ways to integrate conservation and agricultural production. We demonstrate that it ***is*** possible to conserve biodiversity on farms and maintain a productive agricultural enterprise at the same time.

Of course, much can be learned from farmers and other land managers – people with a direct connection to their land. The knowledge we have gained from working with hundreds of landholders over the past 10 years and more is woven into this book. A lot of our existing work is continuing and new projects have just commenced. Therefore we hope to update and improve this book in five to 10 years time and we welcome any feedback.

The authors
October 2010

Acknowledgements

It was possible to write this book only because of the generous support of the now sadly defunct organisation Land & Water Australia, which provided a Senior Fellowship to David Lindenmayer in 2009. We hope this once-great institution will be replaced by something even better and that future governments do not lose sight of the importance of conserving Australia's special biodiversity.

The research and management insights reported in this book have come, in part, from more than a decade of work in south-eastern Australia. That work has drawn on research funds from many sources since 1997. Supporting organisations include the Australian Government Department of Environment, Water, Heritage and the Arts' Box-Gum Grassy Woodland Stewardship Program, the Murray Catchment Management Authority, the Lachlan Catchment Management Authority, the Murrumbidgee Catchment Management Authority, the Commonwealth Environmental Research Facility (CERF)'s Applied Environmental Decision Analysis (AEDA) hub, the Australian Research Council, Land & Water Australia, Rural Industries Research and Development Corporation, the ACT Government, the Natural Heritage Trust, the Caring for Our Country grants scheme (in collaboration with the Victorian Department of Sustainability and Environment, the Murray Catchment Management Authority and the North East Catchment Management Authority and the Goulburn Broken Catchment Management Authority), the Thomas Foundation, the former CSIRO Divisions of Entomology and Sustainable Ecosystems, the New South Wales Government Department of Environment, Climate Change and Water and the New South Wales Roads and Traffic Authority.

Our research on temperate woodlands and replanted native vegetation would not be possible without the support of a very large number of highly dedicated landholders who allow access to their farms for our ongoing studies. Nor would it have been possible to complete much of the work on the birds of woodlands and plantings without the extraordinary contributions of the dedicated volunteers from the Canberra Ornithologists Group, particularly Ian Anderson, Jenny Bounds,

Mike Doyle, Steve Holliday, Bruce Lindenmayer, Noel Luff, Martyn Moffat, Terry Munro, Henry Nix and Peter Roberts.

Merridee Bailey assisted with a range of key tasks that made the initial concept of a book a reality.

We especially thank John Manger from CSIRO Publishing, whose support and encouragement with this project are deeply appreciated.

We also thank a range of colleagues from the Fenner School of Environment and Society at The Australian National University for their outstanding collaborative efforts over the past decade. They include Ross Cunningham, Jeff Wood, Emma Knight, Joern Fischer and David Salt.

1

Introduction and background

Our aim in this book is to highlight some ways to promote wildlife conservation on farms. We are acutely aware that managing land for multiple goals is a difficult task and that not all parts of a farm will be managed in the same way or with the same order of priorities. Given this, we provide new information to help landholders make decisions about ways they might manage parts of their farms. To do this, we describe the characteristics of good remnants, good plantings, good paddocks, good rocky outcrops, good waterways and then, collectively, what makes a good farm for wildlife.

What are temperate woodlands?

Our focus in this book is on the temperate woodland region of south-eastern Australia. Temperate woodlands are vegetation types that occur in temperate Australia and support scattered or widely-spaced trees 10–30 metres tall, with the crowns of the trees shading less than 30% of the ground.[1] In south-eastern Australia, temperate woodlands can generally be thought of as the interface between taller, wetter forested areas on the coast and the drier, hotter grasslands and shrublands of the interior (see Figure 1.1), although there are also many types of coastal and sub-alpine woodlands.

The kinds of temperate woodlands which are the focus of this book run primarily to the west of the Great Dividing Range from southern Queensland, through New South Wales and the Australian Capital Territory, into Victoria, Tasmania and the south-east of South Australia (Figure 1.1). This area coincides strongly with the nation's major wheat–sheep belt where much of the original

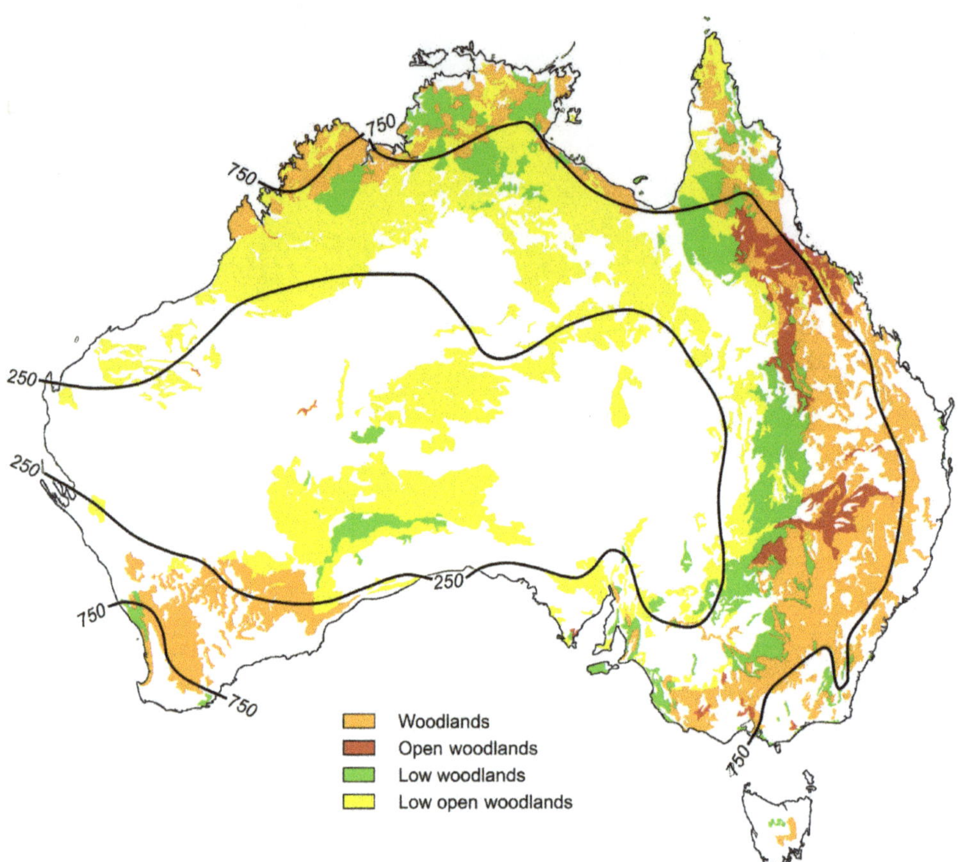

Figure 1.1: The location of Australia's temperate woodlands. (Source: Johnson (2003) based on *Atlas of Australian Resources – Vegetation 1990*, prepared by Troy Honeman and Will Smith)

vegetation cover has been modified extensively by clearing and/or grazing. Most of our work has been based on studies on the South West Slopes of New South Wales, but we anticipate that many of the management suggestions that we make in this book will be relevant to temperate woodland environments elsewhere in south-eastern Australia.

The woodlands of the South West Slopes of New South Wales

A number of different types of woodland communities occur in the South West Slopes, all of which can be identified by the dominant overstorey species. Unique combinations of soils, climate and landscape position determine which plants will flourish at a particular place. One common type of woodland is Box-Gum grassy woodland, a threatened ecological community which is dominated by White Box *Eucalyptus albens*, Blakely's Red Gum *E. blakelyi* and Yellow Box *E. melliodora*.

This community can be found growing on fertile soils in low-lying parts of the landscape (coinciding with prime agricultural areas). The understorey often supports a diverse range of native grasses and wildflowers such as lilies, orchids and everlasting daisies. On heavier clay soils towards the Riverina plains, Box-Gum woodland intergrades with an open woodland dominated by Grey Box *E. microcarpa* and White Cypress Pine *Callitris glaucophylla*, and on the less fertile slopes and ridges, Box-Gum woodland merges with a shrubby type of woodland community dominated by Long-leaved Box *E. goniocalyx*, Red Box *E. polyanthemos* and a variety of small shrubs.

Why is conserving biodiversity on farms beneficial and important?

Biodiversity is the variety of life and includes not only species and their genetic variation, but also the key ecological processes that underpin the functioning of productive ecosystems

Almost all farmers are interested in biodiversity. Of Australia's 150 000 farms, 94% have voluntarily undertaken some form of natural resource management activity, citing increased productivity (89%), increased sustainability (88%), protection of the environment (75%) and improved risk management (64%).[2] Nevertheless, some land managers in Australia consider the conservation of biodiversity to be an impediment - something that stops them doing what they want to do on their land. The maintenance of biodiversity on a farm and throughout agricultural landscapes is a benefit, however, not a disadvantage. An increasing number of people around the world believe that biodiversity conservation needs to be an integral part of ecologically sustainable land management and agricultural production. In the remainder of this section, we explain some of the reasons why conserving biodiversity on farms is beneficial and important. The key benefits we touch on are that:

- biodiversity is critical for the maintenance of important ecological processes in woodland ecosystems
- biodiversity conservation can have financial benefits for farming enterprises
- there are regional, national and global heritage values associated with the conservation of Australian plants and animals
- biodiversity conservation can provide opportunities for improving human lives in the future through medical science.

Setting goals is fundamental to all successful management activities on a farm

Ecosystem function benefits

Ecosystems provide the fundamental resources on which humans depend - food and fibre, water and clean air.[5]

Box 1.1. What is biodiversity? And why is it important for farmers?

Biodiversity can be defined in many different ways. In fact, the scientific literature contains almost 100 different definitions.[3] Put simply, biodiversity is the variety of life on earth. This means that biodiversity encompasses not only the number of species in the environment but also the genetic variability characterising each species. Many definitions of biodiversity also include: (1) the variability in patterns of species throughout landscapes; and (2) the ecological processes, such as seed dispersal, pollination and nutrient cycling, that underpin how well ecosystems function.

The maintenance of biodiversity is critical for productive and profitable farms

The supply and quality of many of these resources depend on maintaining key ecosystem functions or processes such as pest control, seed dispersal, pollination, the decomposition of waste, and nutrient and water cycling.[5] Without biodiversity these essential ecological processes are either severely impaired or do not happen. These essential processes also function more effectively when many different species are present and interacting with each other, thereby reducing the risk of these processes failing entirely if one or more individual species declines or becomes extinct. In this sense a diversity of biota acts as a kind of environmental insurance policy (see Figure 1.2).[6, 7] Good environmental management is therefore intimately linked to the maintenance of productive and profitable ecosystems *and* the maintenance of biodiversity (see Box 1.2).

Financial benefits

There are compelling financial reasons why the conservation of biodiversity on farms is important. On a global basis, it has been estimated that the environment returns more than $US33 trillion in goods and services to society each year – about 1.8 times global GDP.[11] Similarly, investment in intact ecosystems has been found to have a cost-benefit ratio of 1:100, meaning that for every $1 invested in an intact ecosystem, benefits worth $100 are reaped over time. The rivers, wetlands and floodplains of the Murray–Darling Basin are estimated to provide $A187 billion in ecosystem services annually.

At a farm, catchment, state and federal level, governments are recognising the importance of maintaining biodiversity. Increasingly, financial incentive schemes are being developed to assist landholders to conserve biodiversity on their land. Catchment Management Authorities (CMAs) run incentive schemes for farmers across Australia. The BushTender program in Victoria is an example at the state government level. The Box-Gum Grassy Woodland Stewardship Program[12] is an equivalent national scheme. These schemes pay landholders to manage parts of their properties to achieve better conservation outcomes including improving the

Figure 1.2: Examples of species influencing key ecosystem function. (a) A colony of Sugar Gliders can consume more than 20 kilos of insects each year, including invertebrates that would otherwise contribute to paddock tree dieback. (Photo by Esther Beaton). (b) Australian White Ibis – a bird species known to consume large quantities of pasture insects. (Photo by Julian Robinson). (c) Lesser Long-eared Bats. A wide range of species of microbats occur in agricultural landscapes. A single individual can consume large quantities of pest insects – sometimes more than half its body weight in just one night of foraging. (Photo by Mason Crane)

Financial incentives programs now exist that pay landholders to conserve native plants and animals on their farms

condition of native vegetation. Some of these schemes, such as the Box-Gum Grassy Woodland Stewardship Program, are being expanded with plans to implement them through large parts of rural Australia and across many kinds of ecosystems. Landholders currently not in these schemes, but who manage their properties with joint conservation and production goals, are more likely to be among those targeted to be paid under such programs in the future.

Other kinds of incentive schemes may develop in the coming years. For example, it is clear that tackling climate change is one of the major social and environmental challenges of this century. Deep cuts in greenhouse gas emissions will be critical, but so will sequestration of carbon. The establishment of areas of native vegetation (including the establishment of native pastures) will be important, and key places to do that include areas of agricultural land where

Box 1.2. The critical need for management objectives for woodland remnants

The array of entities that comprise biodiversity (see Box 1.1) clearly make it a very complex concept and, in turn, something very difficult for most people (including many conservation scientists) to comprehend fully. A recurring theme in the remainder of this book is that there can be a range of factors that can threaten biodiversity. Different kinds of management interventions will be needed to reduce the impacts of these different threatening processes. These management interventions can include the use of fire, controlling grazing, increasing natural regeneration of trees, reducing the occurrence of exotic species, and reducing levels of nutrients in the soil. It is impossible, however, to manage appropriately for everything on every hectare of a farm. In addition, a particular management action, for example, establishing a narrow strip planting, might benefit some species but disadvantage others.[4] It is therefore important to have an explicit set of goals for any intended management action. What is the objective? That is, what do you want to achieve through a given management activity? Why is that goal important? Is it aimed at conserving a particular species or set of species or is it aimed at restoring a particular key ecological process, such as rectifying problems with rising water tables? What are the priorities, that is, what needs to be done first, second, etc? What is the sequence of steps needed to achieve my objectives? What measures should I use to assess the success of my actions? How often will I measure my progress? What will I do if my objectives are not being met? These questions might sound trivial, but it is surprising how often some kind of management action is instigated without consideration of priorities or what is hoped to be achieved by managing a particular area or patch of bush.

extensive clearing has removed native woodland and grassland.[13] Indeed, in 2008, the global carbon market was estimated to be worth $A64 billion annually.[14] A reasonable price of carbon could make it possible to revegetate parts of a farm in ways that will be financially attractive for many landholders. As we show in Chapter 3, well planned and well managed revegetation programs can also produce significant conservation benefits for many species, including a number of declining or threatened ones.[15, 16]

Heritage reasons

Australia's biodiversity is by far the most distinctive resource that makes this nation different from every other country and continent on this planet. Native wildlife are used as emblems to characterise quintessentially Australian icons ranging from football teams to airlines. In addition, Australia's temperate woodlands are environments that are deeply embedded in the nation's history,

Box 1.3. The link between biodiversity and ecological function

Insects are an incredibly diverse group of animals that, by performing different ecological functions, provides valuable services to agriculture. The following three examples illustrate how a diverse range of insects contributes to a diverse range of ecological functions, all of which are critical to maintaining healthy ecosystems.

1. Pollination

Pollination of crop monocultures such as canola in Australia relies heavily on the European Honey Bee *Apis mellifera*. Native insects, such as syrphid flies, can also pollinate some crops. Importantly, high-quality remnant vegetation provides refuge for insect pollinators.[8]

2. Decomposition

A diverse range of insects is involved with decomposition of dead plants and animals. For example, dung beetles break up dung and bury it underground for their larvae to eat; carrion beetles eat and break down dead animals (Figure 1.3a).

3. Predation of pests

Native vegetation supports a diverse range of insect predators, such as lady beetles (Figure 1.3b), that eat aphids and scale insect pests that attack and damage crops. Studies have shown that a diverse range of insect predators can be effective in suppressing insect herbivores that attack crops.[9] In Australia, the value of native vegetation is increasingly recognised as providing habitat for natural predators that can assist in the control of crop pests.[10]

Figure 1.3: Native vegetation supports a diverse range of insect pollinators, decomposers and predators. (a) Carrion Beetle *Ptomaphila perlata*. (Photo by Jessica Griffiths). (b) A Lady Beetle larva. (Photo © CSIRO)

Figure 1.4: The Koala – a species with remnant populations in some areas of temperate woodland in eastern Australia. This iconic species is clearly linked with Australia. But it is also worth a fortune to the country as it attracts over $1 billion annually through the tourism industry. (Photo by Julian Robinson)

heritage and culture. The vast majority of species on this continent can be found in Australia and nowhere else. This is true for virtually every group of plants and animals, from mosses and lichens to flowering shrubs and trees, and from invertebrates to frogs, reptiles, birds and mammals.[17] Australians therefore have a custodial responsibility to conserve species that are unique to this nation. Moreover, we have a custodial responsibility to the Indigenous people who cared for temperate woodlands and their biodiversity for tens of thousands of years before the arrival of Europeans.

Australia's temperate woodlands are also culturally important for Australians of European descent. They are places where gold was first discovered, were home to some of the nation's most notorious bushrangers, are featured in classic paintings by masters such as Tom Roberts, Arthur Streeton, Jane Sutherland and Fredrick McCubbin (see Figure 1.5), and were the backdrop for poems by iconic authors such as Banjo Patterson and Henry Lawson. There are therefore compelling historical and cultural reasons for informed management and

Figure 1.5: (a) Frederick McCubbin, Australia 1855–1917. *Lost* 1886. Oil on canvas, 115.8 × 73.7 cm. National Gallery of Victoria, Melbourne, Felton Bequest, 1940. (b) Arthur Streeton, Australia 1867–1943. *Land of the Golden Fleece* 1926. Oil on canvas 50.7 × 75.5 cm. National Gallery of Australia, Canberra, The Oscar Paul Collection, Gift of Henriette von Dallwitz and of Richard Paul in honour of his father 1965.

Figure 1.6: The blood-clotting proteins in Tiger Snake venom are being examined to develop new drugs to stop excessive bleeding. (Photo by David Lindenmayer)

conservation of temperate woodland environments. The presence of native plants and animals is one of the reasons many people want to live in rural Australia. They experience feelings of well-being through knowing that native plants and animals exist on their properties.

Medical science benefits

Many studies have demonstrated that a loss of biodiversity can lead to increased incidence of human diseases[18] ranging from malaria to Hendra and Australian bat lyssavirus.

There are also many potential medical benefits of Australian biodiversity that are currently unrealised. For example, Australia's snakes are renowned for the efficiency with which they can immobilise their prey by injecting venom. Australian snake venoms may have many uses in human medicine, including pain relief, bacteria control and treatments of digestive problems, cardiovascular disease, stroke and cancer.[19] These medical breakthroughs can only happen if we conserve snakes and their habitats.

Background material for this book

This book is based largely on our collective new insights into biodiversity research and management in Australia's temperate woodlands. It draws substantially upon an array of research projects that have been completed or are ongoing in the Fenner School of Environment and Society at The Australian National University. The overarching aim of these projects is to improve the ecological sustainability of farming enterprises (including the conservation of farm biodiversity).

Our audience

Our intended audience for this book is broad and we hope that many kinds of people with a diverse array of interests will find it useful and informative. Our aim has been to write material that will be valuable for:

- farmers on grazing and cropping properties
- private landholders, including hobby farmers who live close to large urban areas
- winegrowers, who often have valuable areas of woodland near their vineyards
- natural resource managers, such as those associated with Catchment Management Authorities, Landcare groups and state government agencies
- policy makers in state and federal government departments
- members of the lay public with interests in wildlife conservation and management.

We are aware that family farms are increasingly being taken over by large pastoral companies and we hope that the managers of these organisations might be interested in the conservation and management insights in this book. Finally, we also hope that some of our scientific colleagues might find new perspectives in this book that will be of interest to them.

Because our target audience is diverse, we have tried to write this book in an open and engaging style. We have elected to keep the book as short as possible and have not exhaustively summarised the rapidly expanding literature on farm wildlife. We apologise to our scientific colleagues who find this approach annoying and 'unscientific' – our intention was not to offend. We have used a numbering system to refer to appropriate scientific publications which are listed at the end of each chapter.

What is the difference between this book and previous ones?

We dedicated substantial effort to communicating the importance of biodiversity conservation on farms in two previous books – *Wildlife on Farms: How to Conserve Native Animals* (CSIRO Publishing, 2003) and *Woodlands – A Disappearing Landscape* (CSIRO Publishing, 2005). This new book differs in some important ways. In particular, the past two books were based on general conservation and management principles, whereas this volume is underpinned by more than a decade of new research insights and findings. Many of these new perspectives were unexpected or surprising to us when we first discovered them. The structure of this book is also quite different, focusing on the features and management activities that make a good remnant, a good planting, a good paddock, a good rocky outcrop and a good waterway.

The structure of this book

We have written this book as a series of stand-alone chapters, each of which focuses on a different kind of key environmental asset that occurs on a farm. We have done this because we recognise that some people will wish to read only one or two chapters that are directly relevant to them. For example, some land managers will have no rocky areas or perhaps no remnant native vegetation on their farm and their interest may be primarily on how best to establish plantings on their property. In other cases, there will be readers for which all chapters might be useful and who might want to read the chapters in sequence. We do note that some of the key themes in this book are relevant to many chapters and this has resulted in some unavoidable repetition.

The content of the following chapters

Chapter 2 focuses on remnant native vegetation, and particularly temperate woodland. We examine the different attributes of woodland remnants and which ones are important for particular kinds of native animals and plants.

The title of Chapter 3 is 'What makes a good replanting?' That chapter discusses where and what to replant to restore areas so they are effective for biodiversity. Chapter 3 is important because considerable effort over the past 20 years has been targeted at replanting native vegetation.

Chapter 4 is about paddocks. It examines the characteristics that can be maintained or created to make paddocks valuable for biodiversity. The material we cover in Chapter 4 includes a discussion of native pastures, paddock trees, and fallen timber.

Chapter 5 is a discussion of the conservation and management of rocky outcrops. Rocky outcrops can be one of the most species-rich environments on a farm, even though their important roles in conserving biodiversity are often either overlooked or unrecognised. Reptiles are a particular focus of that chapter because rocky outcrops are an especially important environment for them.

Chapter 6 focuses on waterways. High levels of species richness of many groups of plants and animals are often associated with streams, wetlands, farm dams and irrigation channels. These are therefore critical parts of farms. We readily admit that we have completed less direct research work on these parts of a farm than, for example, plantings, remnant native woodland, rocky outcrops and paddocks (but see [20–22]).

Chapter 7 highlights how each of the broad kinds of environmental assets on a farm can collectively contribute to overall farm-level biodiversity. It also discusses

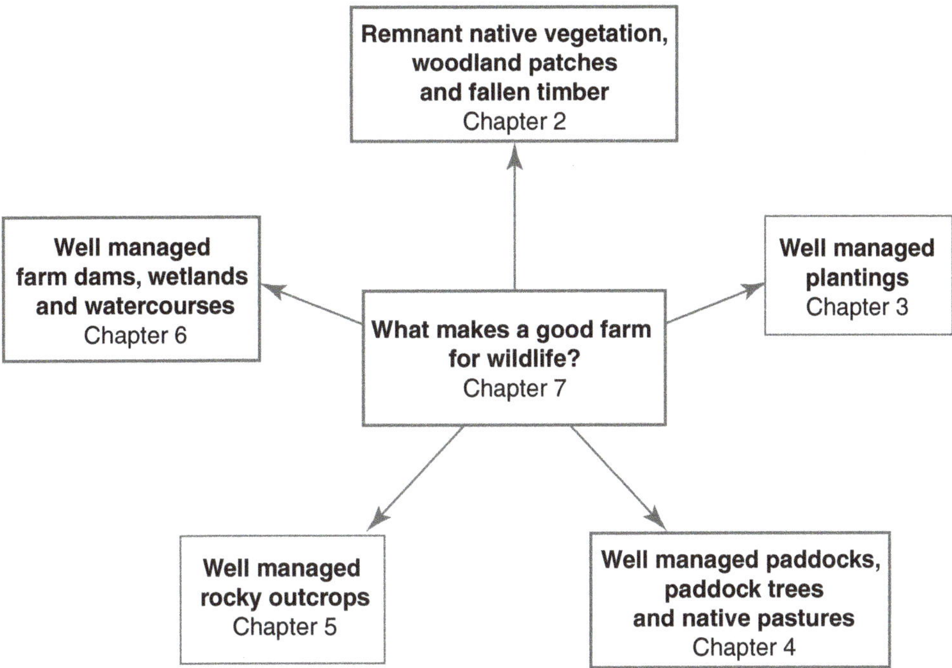

Figure 1.7: Making a good farm for wildlife requires consideration of all of the key environmental assets that occur on a farm, each of which is the focus of a different chapter in this book.

ways to maintain or improve farms to better integrate conservation with agricultural production.

References

1. Hobbs, R.J. and Yates, C.J., eds. 2000. *Temperate Eucalypt Woodlands in Australia: Biology, Conservation, Management and Restoration.* Surrey Beatty and Sons: Chipping Norton.
2. Australian Bureau of Statistics. 2008. 'Natural resource management on Australian farms 2006–07'. Australian Bureau of Statistics: Canberra.
3. Bunnell, F. 1999. What habitat is an island? In *Forest Wildlife and Fragmentation: Management Implications.* (Eds Rochelle, J.A., Lehmann, L.A., and Wisniewski, J.) pp. 1–31. Brill: Leiden, Germany.
4. Cunningham, R.B., Lindenmayer, D.B., Crane, M., Michael, D.R., MacGregor, C., Montague-Drake, R. and Fischer, J. 2008. The combined effects of remnant vegetation and tree planting on farmland birds. *Conservation Biology* **22**: 742–752.
5. Daily, G.C., ed. 1997. *Nature's Services: Societal Dependence on Natural Ecosystems.* Island Press: Washington, DC.

6. May, R.M. 2001. *Stability and Complexity in Model Ecosystems.* Princeton University Press: Princeton, New Jersey.
7. Walker, B.H. and Salt, D.A. 2006. *Resilience Thinking.* Island Press: Washington, DC.
8. Cunningham, S.A., FitzGibbon, F. and Heard, T.A. 2002. The future of pollinators for Australian agriculture. *Australian Journal of Agricultural Research* **53**: 893–900.
9. Letourneau, D.K., Jedlicka, J.A., Bothwell, S.G. and Moreno, C.R. 2009. Effects of natural enemy biodiversity on the suppression of arthropod herbivores in terrestrial ecosystems. *Annual Review of Ecology, Evolution, and Systematics* **40**: 573–592.
10. Holloway, J.C., Furlong, M.J. and Bowden, P.I. 2008. Management of beneficial invertebrates and their potential role in integrated pest management for Australian grain systems. *Australian Journal of Experimental Agriculture* **48**: 1531–1542.
11. Costanza, R., Darge, R., Degroot, R., Farber, S., Grasso, M., Hannon, B., Limburg, K., Naeem, S., Oneill, R.V., Paruelo, J., Raskin, R.G., Sutton, P. and Vandenbelt, M. 1997. The value of the world's ecosystem services and natural capital. *Nature* **387**: 253–260.
12. Department of the Environment, Water, Heritage and the Arts. 2009. 'Environmental Stewardship Program. Box-Gum Grassy Woodland Project.' Department of the Environment, Water, Heritage and the Arts: Canberra.
13. Bekessy, S.A. and Wintle, B.A. 2008. Using carbon investment to grow the biodiversity bank. *Conservation Biology* **22**: 510–513.
14. Capoor, K. and Ambrosi, P. 2008. 'State and trends of the carbon market 2008'. World Bank, Washington DC. http://siteresources.worldbank.org/NEWS/Resources/State&Trendsformatted06May10pm.pdf
15. Munro, N., Lindenmayer, D.B. and Fischer, J. 2007. Faunal response to revegetation in agricultural areas of Australia: a review. *Ecological Management and Restoration* **8**: 199–207.
16. Lindenmayer, D.B., Knight, E.J., Crane, M.J., Montague-Drake, R., Michael, D.R. and MacGregor, C.I. 2010. What makes an effective restoration planting for woodland birds? *Biological Conservation* **143**: 289–301.
17. Lindenmayer, D.B. and Burgman, M.A. 2005. *Practical Conservation Biology.* CSIRO Publishing: Melbourne.
18. Chivian, E. and Bernstein, A. 2008. *Sustaining Life: How Human Health Depends on Biodiversity.* Oxford University Press: Oxford, UK.
19. Stocker, K.F., Todd, P.W. and Bier, M., eds. 1990. *Medical Use of Snake Venom Proteins.* CRC Press, Inc: Boca Raton, Florida.
20. Hazell, D., Cunningham, R., Lindenmayer, D., Mackey, B. and Osborne, W. 2001. Use of farm dams as frog habitat in an Australian agricultural landscape:

factors affecting species richness and distribution. *Biological Conservation* **102**: 155–169.

21. Hazell, D., Hero, J.M., Lindenmayer, D.B. and Cunningham, R.B. 2004. A comparison of constructed and natural habitat for frog conservation in an Australian agricultural landscape. *Biological Conservation* **119**: 61–71.
22. Hazell, D., Osborne, W. and Lindenmayer, D.B. 2003. Impact of post-European stream change on frog habitat: south-eastern Australia. *Biodiversity and Conservation* **12**: 301–320.

2

What makes a good remnant?

In a nutshell

A good temperate woodland remnant will typically have several or all of the following features:

- large diameter living and dead overstorey trees
- trees belonging to several age classes so large trees are replaced as they age and collapse
- the presence of younger trees, indicating that natural regeneration is taking place
- strategies (e.g. fencing) to limit the risk of overgrazing and excessive trampling by stock
- a well developed understorey – for those woodland types that naturally support understorey trees and shrubs
- a ground cover dominated by native plants, including native grasses
- a ground layer characterised by fallen timber, a well-developed layer of leaf litter and an intact surface crust of mosses and lichens in grassless areas.

As we stress throughout this chapter, if a remnant does not have some of these features it does not mean it is without value for biodiversity. Rather, there will be ways to improve it.

Introduction

This chapter is about remnant native vegetation, primarily temperate native woodland. Remnant native vegetation encompasses not only overstorey trees but

All of the natural layers of woodland remnants are important – the ground cover, the understorey and the overstorey trees

also the understorey and ground layers. A woodland remnant can be a large patch of trees (plus understorey and ground cover) extending over tens or even hundreds of hectares or it can be as small as a single paddock tree.

Remnant native vegetation is critical for the persistence of biodiversity: for birds it is three times more important in terms of the number of species it supports than planted vegetation and it also supports a wider array of declining woodland bird species.[1]

In this chapter we discuss some of the key features of a good remnant and how these features influence wildlife. These features include vegetation condition, patch size, and adjacency to other areas of vegetation (including plantings; see Chapter 3). We discuss native woodland remnants by their various vertical and horizontal layers and structural classes, and consider why each one is important for biodiversity using examples from our research over the past decade to illustrate key topics. We outline management actions that can maintain existing areas of good remnant vegetation or improve them for biodiversity. Although our focus is on what we consider to be a 'good' remnant, we stress that **any** remnant native vegetation on a farm is good for animal and plant biodiversity because it will support at least some native species.

Most of this chapter explores issues associated with patches of remnant native vegetation. Scattered paddock trees and native pastures are also important elements of remnant native vegetation, but we discuss them in Chapter 4.

Vegetation condition

For the purposes of this book, vegetation condition is measured as the degree to which a remnant varies from 'reference sites' that are relatively unmodified stands of the same vegetation community.[2] The reference concept is based on the assumption that ecosystems approaching conditions that prevailed prior to major periods of modification (e.g. European settlement) will generally reflect the conditions to which native plant and animal communities are best adapted.[3]

Attributes that are used to assess vegetation condition in the context of biodiversity are those elements of remnant native vegetation that represent important habitat for wildlife and are indicative of ecosystem function. These can include (among many things) the number of large hollow trees, the volume of fallen timber, the cover of understorey shrubs, and the presence of regeneration.

It is important to assess the condition of a particular remnant relative to an equivalent vegetation community. For example, if good vegetation condition was based on the amount of shrub cover in a remnant, then naturally shrubby vegetation communities such as Red (or Mugga) Ironbark forest would always be considered in better condition than naturally grassy vegetation communities such as Yellow Box woodland.

In the following section we discuss key attributes of vegetation condition for the ground, understorey and overstorey layers of remnants as well as their value for biodiversity. We also outline the processes that can degrade vegetation condition and how these threats can be managed and reversed.

The ground layer

The ground layer of well-managed temperate woodland remnants can support many species of native plants, including orchids, lilies, forbs and native grasses. Other key components of the ground layer include leaf litter, small patches of open bare ground, moss and lichen cover, surface rocks, and fallen timber. Each of these components has been found to be important in the habitat requirements of particular woodland fauna, including several which are declining or threatened and of conservation concern (Table 2.1). The nature and condition of the ground layer of a woodland remnant affects whether the site can offer various types of shelter sites, food sources, or other resources such as materials for nest construction to animals.

Threats to the ground layer and their management

Several factors can degrade the ground layer of temperate woodland remnants and reduce the suitability of habitats for biodiversity. These include overgrazing

Table 2.1. Species for which different components of ground layer have been shown to be important.

Attribute	Example of responding species	Ecological role
Leaf litter	Jacky Winter, Hooded Robin, overall richness of woodland-specialist birds[4, 5]	Leaf litter provides habitat for insect prey for woodland birds
Moss and lichen cover	Restless Flycatcher, Hooded Robin, Dusky Woodswallow[4]	Mosses and lichens are used by many species of birds in nest construction
Surface rock	Reptile diversity[6]	Rocky environments are home to a large number of species of reptiles, many of which are virtually restricted to these areas
Fallen timber	Common Ringtail Possum, Brown Treecreeper, overall beetle diversity[4, 7, 8]	Fallen timber provides nesting and sheltering places for many animals ranging from the Common Ringtail Possum to numerous species of beetles. The Brown Treecreeper also uses fallen timber for foraging
Native grass cover	Southern Rainbow Skink, Superb Parrot[4, 7]	Native grasslands are key habitats for many animals and provide a place where they can find food

Figure 2.1: Two of many species which are known to respond strongly to different attributes of the ground layer: (a) Hooded Robin. (Photo by Julian Robinson). (b) Southern Rainbow Skink. (Photo by Damian Michael)

Figure 2.2: Examples of the amazing variety of shapes and colours of beetles found in leaf litter under Blakely's Red Gum and Yellow Box trees. (Photo by Philip Barton)

Box 2.1. Leaf litter and invertebrates

Tree litter varies from tree to tree, and is a vital part of the ground layer. It includes leaves, sticks, bark and dropped flowers and seeds from the canopy above.[9] Many hundreds of species of microbes, fungi and invertebrates live in litter, breaking it down and incorporating it into the soil. The larger invertebrates in tree litter are also an important food source for reptiles, birds and small native mammals. Removing litter takes away these attributes from a landscape. This means less biodiversity and less food for birds, reptiles and mammals. The number of beetle species that can be found in the leaf litter under *Eucalyptus* trees is amazing (Figure 2.2), and the differences in leaf litter under each eucalypt species can affect what species are found. In a study of beetles under Blakely's Red Gum and Yellow Box trees, over 150 species of beetle were identified, with some species found under one eucalypt species, but not the other.[10] All of these beetles are performing different ecological functions: some are eating dead leaves and bark, some are preying on other invertebrates, and others are eating fungus.

and trampling by domestic livestock, browsing by rabbits and kangaroos, introduction of nutrients from fertiliser and stock camps, weed invasion, over-abundance of exotic annual grasses, firewood harvesting, bush-rock removal and raking leaf litter.

Limiting overgrazing and trampling by livestock, and reducing fertiliser application are two important ways to manage native ground cover

Overgrazing can lead to the loss of native orchids, lilies, forbs (non-woody plants that are not grasses, sedges or rushes) and cryptogams (plants that reproduce by spores) from the ground layer.[3, 11, 12] It can reduce the amount of native grass tussock cover which is a key habitat component for a range of reptiles[7] and plays key functional roles in ecosystems such as maintaining soil surface condition and locking up soil nitrates thus reducing annual weeds.[13, 14] Temperate woodlands with an intensively grazed ground layer also appear to be those most attractive to the Noisy Miner – a native honeyeater that aggressively excludes many other species of native birds (see Box 2.7).[15]

The widespread use of fertiliser and excessive nutrient enrichment from animal dung represents another threat to the integrity of the ground layer of temperate native woodlands. Areas subject to treatment with large amounts of fertiliser may experience increased exotic pasture growth. The ground cover of many (although not all) native plants can be significantly reduced, however, and some native plant species can be lost. In contrast, exotic plant species cover and exotic species richness is increased.[16–19] An over-abundance of exotic annual grasses can significantly impair habitat suitability for some native species of conservation concern. For example, the Eastern Yellow Robin is rare in areas with extensive

Figure 2.3: An area of former woodland that has been extensively cleared and degraded. (Photo by Philip Gibbons)

amounts of exotic annual grass cover. Conversely, where the ground layer is dominated by native grass cover, species like the threatened Superb Parrot are more often found and the diversity of reptiles is higher.[4, 7]

Some of the threats listed above can be addressed with good farm management. For example, the intensity of grazing pressure can be controlled by fencing,[20] although the construction of fences needs careful consideration. Barbed wire on the top strand of a fence should be avoided wherever possible because many kinds of flying and gliding animals can become entangled in it (Figure 2.4). For those fences where it is not feasible to substitute barbed wire with straight wire, it can be useful to enclose barbed wire in a poly-pipe covering or to tie flagging tape to wire to make it more visible to flying and gliding animals.

Changing from traditional set-stock grazing to rotational or cell grazing methods is another significant way a land manager can improve the ground layer of patches of remnant native woodland. Reducing grazing pressure by domestic livestock and using strategic grazing at certain times (such as in late winter–early spring) can limit growth and flowering of exotic annual grasses. This will help promote the development of swards of native grasses (see below), many kinds of native ground cover plants, and moss and lichen mats.

The use of prescribed fire also may help reduce high levels of phosphorus in the soil.[16] Only an appropriate fire regime or sequence of fires should be applied, however. Burns that are too frequent may over-simplify the structure of the understorey layer of some woodland types with the risk of negative impacts for animals associated with it. In some novel research, sugar was successfully used to reduce soil nitrate and restore native perennial grass (see Box 2.2). Direct seeding of other kinds of native ground cover plants also has been found to be successful in some regions, provided the initial exotic ground cover is controlled or removed.[21]

Halting practices like raking of leaf litter and fine woody debris is important for maintaining the integrity of the litter layer and promoting habitat suitability for

Figure 2.4: Animals tangled in barbed wire. (a) Fruit Bat. (Photo by David Lindenmayer). (b) Magpie. (Photo by Rodney van der Ree). (c) Squirrel Glider. (Photo by Rodney van der Ree). (d) Tube-nosed Bat. (Photo by Rodney van der Ree)

declining or threatened species, such as the Hooded Robin and the Jacky Winter. Controlling firewood collection will benefit the numerous species of plants and animals dependent on fallen timber.[22] Finally, bush-rock removal should be prevented. By not removing bush-rocks, rocky areas can continue to have significant positive effects on the diverse array of reptile species associated with the ground layer (see Chapter 5).[6]

Box 2.2. A sweet end to weeds

Exotic annual plants thrive under higher levels of soil nitrogen than typically occurred in temperate woodlands prior to European settlement. In a novel experiment,[17] applications of carbohydrate in the form of sugar were used to reduce soil nitrate. The sugar increased activity among soil micro-organisms that use nitrogen. This created more favourable conditions for native perennial grasses to compete with exotic annuals. Sites treated with sugar and seeded with Kangaroo Grass were successfully revegetated with a sward of perennial native grasses.

The understorey layer

An understorey is typically defined as the vegetation layer between the ground layer (up to approximately one metre) and the overstorey. Some woodland vegetation types support only a sparse understorey.[3] Nevertheless, the understorey is important in many kinds of woodland remnants and it can add a vegetation layer that influences the occurrence of species like the Diamond Firetail (see Table 2.2).[4] Key components of the understorey include shrubs, wattles and regenerating eucalypts (regrowth).

Threats to the understorey layer and their management

The understorey layer is absent from many remnant patches of woodland where it once would have naturally occurred. Grazing by domestic livestock and rabbits as well as competition from exotic plants are the principal reasons for this decline. Reducing grazing pressure from stock and feral herbivores like rabbits is one way to facilitate the development of the understorey layer, and can allow natural regeneration of overstorey trees to produce the important early growth stage of woodland (see Box 2.5). Reducing grazing pressure can be achieved by fencing remnants or by employing strategic grazing regimes like rotational grazing or cell grazing.[20, 26, 27]

Table 2.2. Species for which different components of understorey layer have been shown to be important.

Attribute	Example of responding species
Amount of understorey cover	Eastern Yellow Robin, White-browed Scrubwren, Red-browed Finch[4, 5]
Number of medium-sized stems	Eastern Yellow Robin[4]
Number of dead shrubs	Rufous Whistler, Brown Treecreeper[4]
Presence of regrowth trees in the understorey	Ragged Snake-eyed Skink, Marbled Gecko, Eastern Yellow Robin, Jacky Winter, Black-chinned Honeyeater, Brown Thornbill[4, 5, 7]
Number of vegetation layers	Diamond Firetail[4]
Presence of some *Acacia* species	Squirrel Glider[25]

Box 2.3. Is fallen timber on a farm good or bad?

Prior to European settlement, temperate woodlands naturally supported large volumes of fallen logs, sometimes exceeding 200 cubic metres per hectare.[3, 23] As discussed in this chapter, many native species have evolved to use this habitat resource. Some people believe that having fallen timber on a farm is a problem that should be 'cleaned up'. After all, isn't it fuel for fires and a haven for pests like the European Rabbit and the Red Fox? Large logs are not fine fuels that readily ignite and spread fires. Rather, they often act as micro fire-breaks and can limit the spread of fire. It is true that large logs can shelter rabbits and foxes; but addressing this problem requires the destruction of these pests through approaches other than removing logs, such as laying poison baits and ripping warrens (but see Box 5.3 in Chapter 5). Controlling pests like foxes will not only benefit native wildlife, but have benefits like increasing the survival of lambs. Fallen timber can also provide important shelter for stock, particularly ewes and lambs.

Limiting overgrazing by livestock and rabbits is a good way to manage the understorey layer of woodland remnants

Another way to promote the development of an understorey layer is to deliberately establish it by underplanting. We have established an understorey layer of shrubs and trees in a series of remnants on the South West Slopes of New South Wales as part of a partnership project between local landholders, The Australian National University and the Murrumbidgee Catchment Management Authority (see Figure 2.6). This underplanting work was done because previous work had indicated that the absence of a natural understorey was a factor limiting populations of native birds, and a trial program was instigated to test the response of a number of bird species of conservation concern. Such shrubs are best established in thickets, leaving some open areas between the thickets. The experiment has not yet been running long enough for there to be definitive results, but early indications are that there are positive conservation outcomes from this work.

Box 2.4. Wattles and worms

The value of understorey vegetation for many native animal species has been well established. Wattles, however, have many other surprising roles that can influence farm enterprises. Because wattles are nitrogen-fixing plants they can promote the growth of surrounding trees such as eucalypts. A particularly unexpected effect of wattle has emerged from work conducted in Brazil, which has shown that condensed tannins from the Black Wattle significantly reduce the numbers of intestinal parasites in sheep.[24]

Figure 2.5: The Jacky Winter is a bird of conservation concern which responds strongly to the presence of regrowth trees in the understorey. (Photo by Suzi Bond)

Old growth and regrowth are different growth stages of woodland and support different kinds of animals and plants. Both kinds of remnants are needed on a farm.

Finally, it is important to ensure that shrubs and regrowth trees are not cut down and removed after they die. This is because of their value as habitat for native animals, even when dead. For example, declining bird species like the Brown Treecreeper and the Rufous Whistler are most likely to be found in woodland remnants that support standing dead shrubs.[4]

Figure 2.6: (a) An understorey planting in a grassy Box-Gum woodland near Junee on the South West Slopes of New South Wales. (Photo by Damian Michael). (b) Francois and Lyn Retief who have completed extensive understorey plantings on their farm near Ladysmith. (Photo by Sachiko Okada)

Box 2.5. A prickly subject – the staged approach to shrub regeneration and weed removal

Since European settlement of woodland environments, many areas have lost their shrub layer. Shrubs, particularly dense, leafy or prickly shrubs, such as Bursaria, provide excellent habitat for many species of small native birds. In many plantings or remnants, landholders may feel inclined to remove weeds such as Sweet Briar, Blackberry or Tree Lucerne, which is undeniably an excellent long-term goal. Such weedy shrubs, however, can still provide habitat for small birds. For instance, we have regularly observed the Speckled Warbler and Red-capped Robin at sites with a dense cover of Tree Lucerne. When removing such weeds, consider taking a two-phased approach whereby native shrubs are established before, or at least in conjunction with, the removal of the non-native shrubs. Shrubs are best established in thickets, leaving some open areas between the thickets.

The overstorey layer

The overstorey layer is usually the most obvious part of remnant native vegetation and may, in some woodland remnants, be the only remaining layer. Many characteristics of the overstorey may have a considerable influence on the suitability of woodland remnants for biodiversity and a subset of these are listed in Table 2.3. There is also a strong effect of the dominant tree species which characterises a patch of remnant native woodland. For example, different beetle assemblages occur under Yellow Box compared with Blakely's Red Gum[10] (see Box 2.1). Similarly, we found on the New South Wales South West Slopes, the

Figure 2.7: (a) Old-growth woodland remnants. (b) Coppice regrowth woodland. (Photos by David Lindenmayer)

Box 2.6. The growth stages of woodland remnants and the importance of woodland regrowth for biodiversity

The terms 'old growth' and 'regrowth' are often thought of in the context of tall wet forests and their management. But our recent work has indicated that different growth stages are also critical in temperate woodland vegetation. This is because growth stages significantly influence the presence and abundance of many species of woodland animals. Three broad growth stages of woodland remnants which have proven significant in our studies of reptiles, possums and gliders are old growth, coppice regrowth and seedling regrowth.[7] Old-growth woodland is dominated by large trees that are generally more than 100 years old. Coppice regrowth is multi-stemmed regrowth from existing living trees recovering after fire, logging or both. Seedling regrowth is natural regrowth originating from seeds germinating after being dropped by overstorey trees.

In our large study on the South West Slopes of New South Wales, the Common Brushtail Possum and the Common Ringtail Possum were most often recorded in old-growth woodland.[7] In contrast, the Wall Skink was more often found in coppice regrowth and old growth than in seedling regrowth. The Marbled Gecko was more likely to be detected in seedling regrowth than coppice regrowth.

One of our major studies of birds on the South West Slopes of New South Wales showed that nine species of declining and/or threatened woodland birds were more likely to occur in regrowth woodland (coppice or seedling) than old-growth woodland.[4] These included the Black-chinned Honeyeater, Crested Shrike-tit, Diamond Firetail, Eastern Yellow Robin, Jacky Winter, Restless Flycatcher and the Dusky Woodswallow. Other work on woodland birds in the Australian Capital Territory region has also highlighted the importance of eucalypt regrowth for a range of birds including the Brown Thornbill and Buff-rumped Thornbill.[5]

Recognition of the value of different growth stages in woodlands is important for several reasons. A critical one is that areas of regrowth woodland have often been regarded as 'rubbish' country in the past and of limited value for biodiversity. This is clearly not the case. Some species exhibit a stronger preference for regrowth woodland than old growth. Second, because different species select different growth stages as habitat, the conservation of these different kinds of vegetation should be part of any strategy aimed at maintaining populations of the full range of species that might occur on a farm. Third, the different kinds of growth stages of remnant native vegetation form part of the portfolio of vegetation assets that underpin farm-level biodiversity conservation. Finally, vegetation development is a dynamic process: stands that are currently regrowth can, with appropriate management and an absence of intensive disturbances, eventually become old-growth stands. Therefore, the long-term maintenance of old-growth woodland requires the maintenance of existing areas of regrowth that can be recruited to an old-growth stage over time.[7]

Figure 2.8: The Ragged Snake-eyed Skink is a species associated with woodland remnants where regrowth trees are a prominent part of the understorey. (Photo by Damian Michael)

Common Ringtail Possum is more likely to occur in woodland remnants dominated by White Box than other woodland tree species.[7]

The responses to the overstorey attributes by the particular species listed in Table 2.3 make good ecological sense. For example, mistletoe plays a range of roles for many animals,[28] ranging from providing places to shelter and nest to providing places to forage. Conversely, the amount of dieback in overstorey trees has a negative effect on a range of species. Three of the example bird species listed in Table 2.3 respond negatively to increasing amounts of dieback, possibly because such overstorey trees in poor condition do not provide suitable food resources for them. This also may explain why the Squirrel Glider does not like to feed or den in overstorey trees with high levels of dieback.[25]

Table 2.3. Species for which different components of overstorey layer have been shown to be important.

Attribute	Example of responding species
The depth of the canopy	Superb Parrot, Squirrel Glider[4, 29]
The amount of mistletoe	Restless Flycatcher, White-browed Babbler, Brown Treecreeper, Crested Shrike-tit, Black-chinned Honeyeater, Dusky Woodswallow[4]
Amount of dieback	Restless Flycatcher, Eastern Yellow Robin, Black-chinned Honeyeater, Dusky Woodswallow, Squirrel Glider[4, 25, 29]
Presence and size of large trees	Squirrel Glider, diversity of beetles[10, 29]
Presence of hollow-bearing trees	Squirrel Glider, Superb Parrot, Eastern Rosella, Crimson Rosella[5, 29, 30]

Figure 2.9: Two highly charismatic species of vertebrates strongly associated with key attributes of overstorey trees in woodland remnants: (a) Squirrel Glider. (Photo by Esther Beaton). (b) White-browed Babbler. (Photo by Julian Robinson)

Threats to the overstorey and their management

The integrity of the overstorey layer of temperate woodlands can be threatened by several factors, but some of the major ones include tree clearing and overgrazing by domestic livestock. Broad-scale clearing has largely ceased across the temperate woodland belt following the introduction of laws governing native vegetation clearance. Nevertheless, the removal of small stands of remnant vegetation, such as to allow the development of farm infrastructure or enterprises such as cropping, or the felling of individual trees (living or dead) for firewood, remain significant threats in many agricultural areas. Controlling firewood collection is a simple and very positive way to address this major problem (see Box 4.1 in Chapter 4).

Overgrazing can prevent tree regeneration and, in turn, preclude the recruitment of new large overstorey trees to stands of remnant woodland[3, 27, 31] (see Box 2.7). The additional nutrients added to the soil by artificial fertiliser as well as by defecating livestock in heavily grazed remnants also can accelerate dieback and tree death in overstorey trees. Therefore, reducing grazing pressure can limit degradation of the overstorey tree layer and facilitate the development of natural regeneration of regrowth stands.[27, 32] Grazing pressure can be regulated by fencing remnants and/or through rotational grazing or cell

Limiting clearing and reducing overgrazing and trampling are two important ways to manage the overstorey layer of remnant woodlands

Box 2.7. A tree regeneration crisis in temperate woodlands?

Eucalypt regeneration is virtually always present in relatively unmodified woodland remnants.[3] In a study in the wheat–sheep belt of New South Wales, however, only 38% of remnants on freehold land contained eucalypt regeneration.[31] The key factors associated with the occurrence of regeneration were grazing, exotic plant cover and remnant size. Remnants that were large, with light or intermittent grazing by stock and a high cover of understorey native plants were most likely to contain regeneration. If we don't address the regeneration crisis then a considerable area of remnant woodland will disappear across the wheat–sheep belt over the next century. Excluding livestock grazing from small remnants, grazing small remnants with stock intermittently, controlling rabbits or planting seedlings in gaps within and around small remnants are the principal solutions. But these actions must be undertaken at a large scale across the wheat–sheep belt if we are to avert a tree regeneration crisis.

grazing.[20, 26, 27] Reducing the amount of fertiliser that is used in an area also can promote the natural regeneration of overstorey woodland trees.

Remnant size

The value of large remnants

A general theme in conservation science is that bigger patches of native vegetation are better than smaller ones because they typically support more species and larger numbers of individuals of any given species. At least seven interrelated reasons have been proposed to explain this widespread pattern:

- More kinds of habitat can occur in larger patches, thereby providing niches for more species.
- Larger patches have more resources (e.g. food and shelter sites) and a greater range of resources, allowing them to support more individuals of a particular species.
- Individual animals that are moving across a landscape may be more likely to find (and then settle) in larger patches.
- Following major disturbances such as bushfires, larger patches are more likely to contain areas of undisturbed habitat where disturbance-sensitive species can survive.
- Rates of predation of the nests of native birds are lower in large patches, particularly away from the edges of these patches.
- Larger patches have more 'core' area relative to 'edge' area. Patch edges can suffer more disturbances and support poorer habitat for many species.

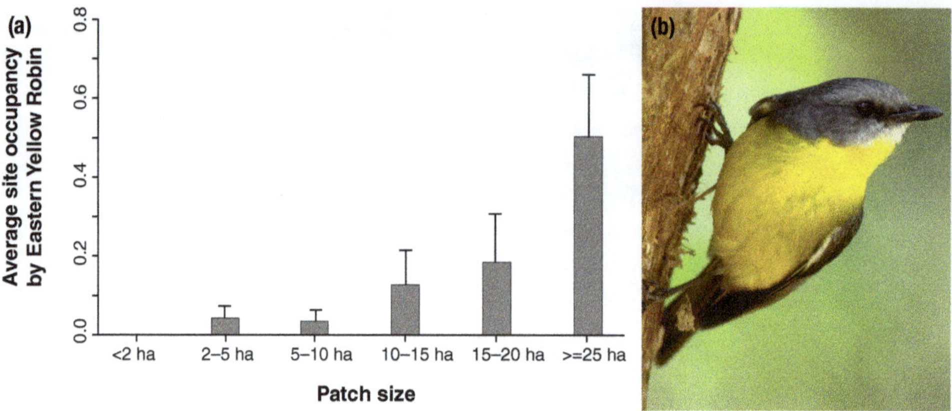

Figure 2.10: (a) Relationship between remnant patch size and site occupancy by the Eastern Yellow Robin. (b) Eastern Yellow Robin. (Photo by Graeme Chapman)

- The effects of the hyper-aggressive Noisy Miner on smaller woodland birds may be less intense in larger patches.

Several of our detailed field studies have demonstrated the importance of large patches of remnant native woodland for biota. For example, declining or threatened woodland birds such as the Hooded Robin, Eastern Yellow Robin and the Brown Treecreeper are more likely to occur in large patches.[4]

The value of small remnants

Large remnants typically represent the best habitat for biodiversity, but small remnants also can be important for conservation. A number of researchers have demonstrated the importance of small remnants for native woodland plant conservation and animal conservation.[33, 34, 35] Even 'patches' as small as an individual paddock tree have been found to be extremely valuable for some animals[30] (see Chapter 4). Sets of small patches of woodland may support many species (and sometimes more species) than a single large patch of equivalent area.[35] This can be because a small patch might be part of a much larger network of patches that together are needed by a range of mobile species.[36] Further, a small, high-quality patch containing particular vegetation attributes that are important for a species of interest (e.g. trees with hollows, or a well-developed understorey layer) might be more important than a larger patch in poor condition that does not contain these features. Small patches are also critical because most of some woodland communities occur as small patches (2 hectares or less).[37]

Larger patches of woodland typically support more species than smaller ones but small remnants are often the priority for conservation on a farm because they are under greater threat of being lost

Box 2.8. Transforming a small remnant into a larger remnant

Given that larger woodland remnants can be better for biodiversity than smaller ones, a key question is: Is there a way to turn smaller remnants into larger ones? One simple way to do this is to establish a fenced area that is larger than the remnant the fence encloses (Figure 2.11 left). Another is to create a single fence that encompasses not only two or more remnants but also the cleared or semi-cleared areas between them (Figure 2.11 right). In this way, trees can regenerate in initially cleared areas and perhaps eventually link previously separated remnants.

Enhancement planting around the edge of an existing remnant or between two existing remnants also can help expand the size of an area of remnant vegetation. We discuss the importance of plantings in the next chapter – What makes a good planting?

Small patches are also critical for conservation because they are typically under greater threat of decline and loss than larger patches. For example, the conservation of large patches across the wheat-sheep belt would not address the tree regeneration crisis in this landscape because it is the smaller remnants that are typically not regenerating and therefore under greater threat of being lost without intervention (see Box 2.7). Similarly, weed problems, dieback and clearing tend to be greater in smaller remnants than they are in larger remnants.

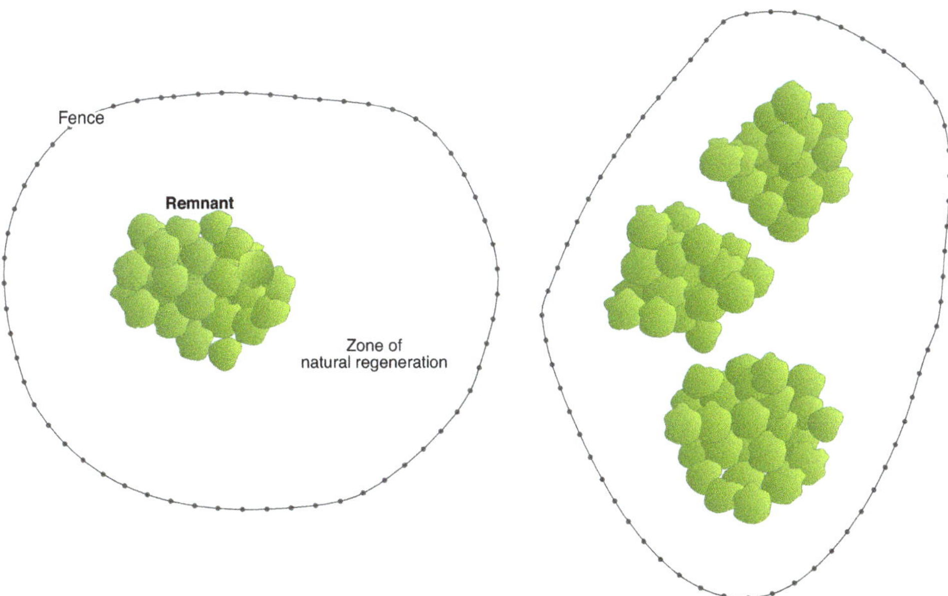

Figure 2.11: Fencing around patches of remnant woodland as well as the cleared areas that surround them to make an existing woodland remnant larger. (Drawing by Clive Hilliker)

Box 2.9. Noisy Miners and woodland remnants

The Noisy Miner is a large and highly aggressive native honeyeater. It has a well-deserved reputation for driving away other native birds from woodland remnants, particularly smaller-bodied species. When this bird is removed from woodland remnants, some of these smaller species quickly return.[38, 39] The insect pest control role of these other bird species is reduced when the Noisy Miner is over-abundant, and this is thought to contribute to tree dieback.[15, 40]

Fortunately, the Noisy Miner is not ubiquitous and populations of this species are larger within woodland remnants in some parts of landscapes than others. Recent work in Victoria suggests that the species is more abundant in the more productive parts of grazing landscapes, such as those on valley floors close to watercourses.[41] Patches of woodland on steep, low productivity areas tend to support lower numbers of the Noisy Miner but higher numbers of smaller species of woodland birds. Similar findings have been made for temperate woodlands in southern New South Wales and show that when the Noisy Miner is absent from higher woodland patches on higher productivity sites, an array of native birds of conservation concern (e.g. the Black-chinned Honeyeater) are significantly more likely to occur. Notably, our other studies have indicated that vegetation structure also can influence the occurrence of the Noisy Miner – the species is typically absent from plantings with a dense understorey layer of wattle trees[42] (see Chapter 3).

Many ecologists believe that the control of the Noisy Miner should be a more widely used form of conservation management on farms – both to improve conditions for woodland birds but also to improve tree health.[40] Once Noisy Miners have been removed from a site, it takes a prolonged period (often years) for a new colony to re-invade.

One option for controlling Noisy Miner numbers is for regional wildlife management agencies to issue destruction permits for pest animals to allow landholders to shoot problem birds that are over-abundant like the Noisy Miner. This method is relatively cost-effective, quicker and more humane that alternative removal methods for the Noisy Miner.[15]

The importance of adjacency and landscape context

Many features of woodland remnants influence their suitability for biodiversity, including the structure and composition of the vegetation, remnant size and the shape of a remnant. Other factors, not of a remnant itself, but of the surrounding landscape can also affect the occurrence of particular species. A prominent example is the amount of native vegetation in the surrounding landscape. There are

The Noisy Miner is an over-abundant native bird

Figure 2.12: Noisy Miners are known for their aggressive exclusion of other bird species. (Photo by Dr Cheng Hiang Lee)

several reasons for this: areas of native vegetation surrounding a remnant may provide suitable additional foraging or nesting areas for a species and increase the chances it can inhabit a remnant. This also may boost the overall numbers of a species in a landscape and reduce the risk of localised extinction occurring. A second reason is that areas of native vegetation surrounding a remnant may contribute positively to the overall connectivity of a landscape or its suitability for the movement and dispersal of a species.[36, 43] Therefore, if a species goes extinct in a remnant, other animals moving through the landscape may help reverse that localised extinction.

The landscape context (or the characteristics of the landscape surrounding a remnant) has a major effect on wildlife occupying a remnant

Landscape context effects have been identified in a number of our studies. For example, the occurrence of birds such as the Black-chinned Honeyeater and the Eastern Yellow Robin is significantly higher within woodland remnants surrounded by other areas of woodland vegetation. The Brown Treecreeper is most likely to be found in those remnants where the surrounding landscape supports many scattered paddock trees. Similarly, the Diamond Firetail is more likely to occur in remnants where the surrounding landscape is dominated by native pastures.[4]

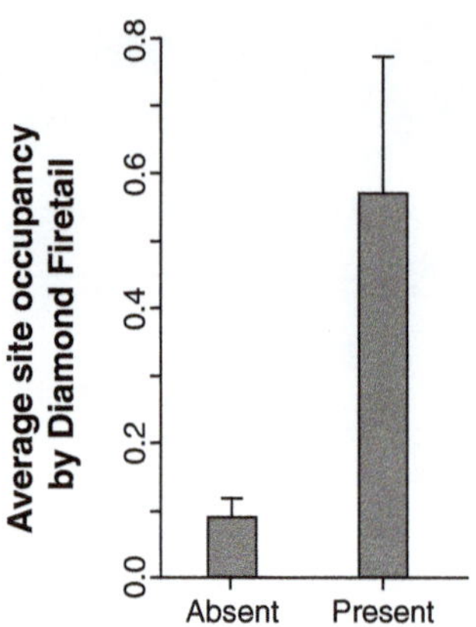

Figure 2.13: Relationship between the occurrence of the Diamond Firetail in a remnant and the amount of native pasture surrounding that remnant.

Summary

Several factors contribute to making a good remnant. These vary from the structure, composition and condition of a patch of remnant vegetation to patch size, patch shape and landscape context. Given an understanding of these key factors, it is then possible to identify management actions to enhance remnant

Box 2.10. An experimental study of the effects of landscape context on the biodiversity of temperate woodland remnants

In 1997, we established a major experimental study to document what happens in patches of remnant woodland where the surrounding landscape is changed from paddocks with scattered trees to a landscape dominated by stands of Radiata Pine. The work is taking place at Nanangroe, near Jugiong in southern New South Wales. The project has involved counting animals in remnant patches before the surrounding plantation establishment commenced, with repeated surveys as the Radiata Pine plantation matured (see Figure 2.14).[44]

Some major changes have occurred in the 13 years since the project started. The bird assemblage has changed from one largely dominated by open country and woodland species to a highly unusual mixture of forest and woodland birds. Many woodland bird species are declining, whereas forest bird species are increasing. Many of the new forest species found in the study were those which inhabited the new stands of pine trees and then spilled over into adjacent woodland remnants.[45]

The key message from the Nanangroe experiment is that landscape context matters. That is, the nature of the landscape surrounding woodland remnants can have a major influence on which species use them.

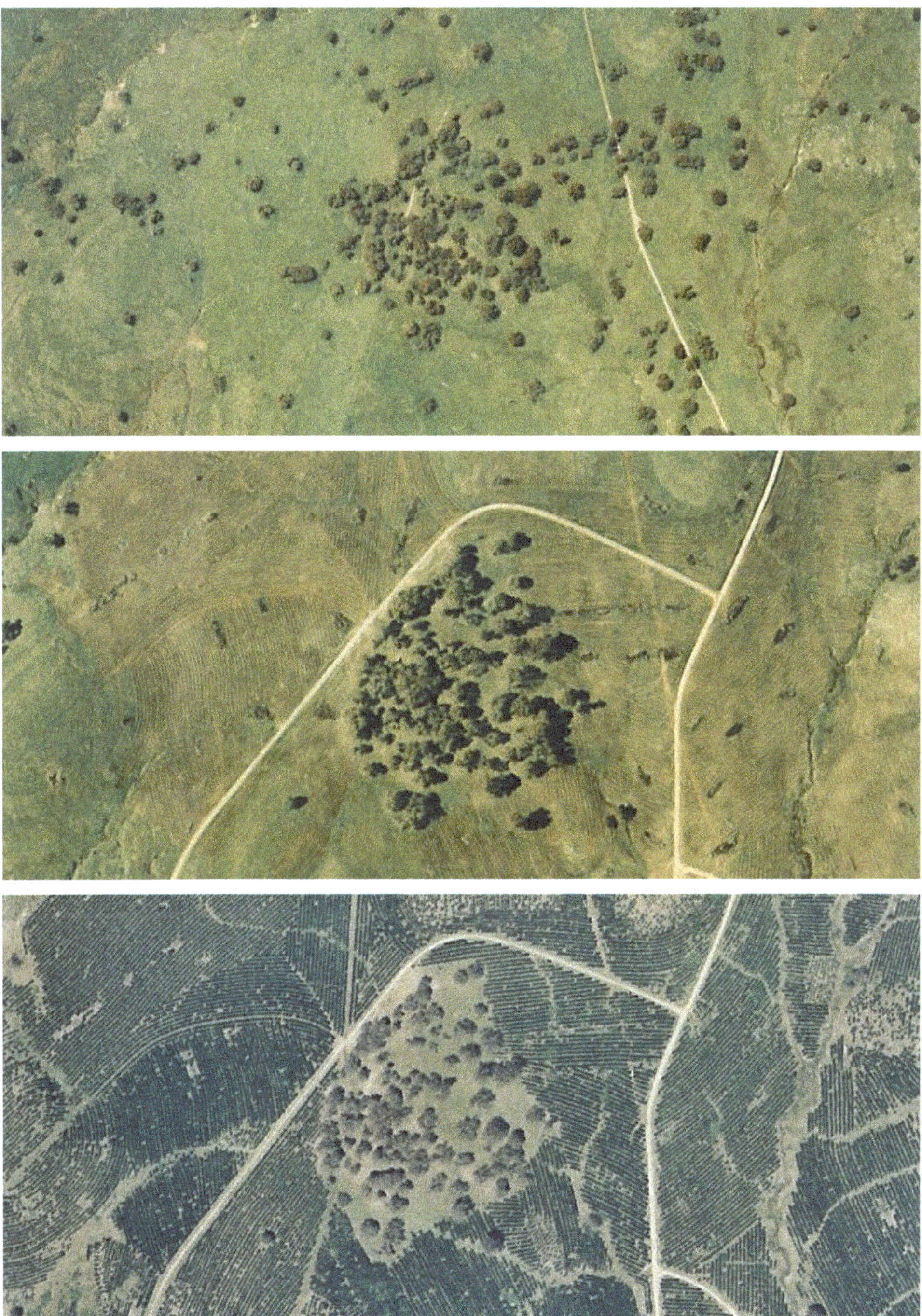

Figure 2.14: Series of changes in the landscape surrounding a patch of remnant woodland at Nanangroe. The dark green areas are stands of maturing Radiata Pine plantation trees. (Photos courtesy of State Forests of New South Wales)

Figure 2.15: The Red-rumped Parrot is a bird species which is lost with the conversion of paddock areas to pine plantations. (Photo by Suzi Bond)

native vegetation for biodiversity. These can include reducing grazing pressure, retaining dead trees, shrubs and surface rocks, making remnants larger (e.g. through enhancement planting or strategic fencing) and controlling weeds.

Woodland remnants can be managed for different purposes and in response to different objectives (see Box 1.2 in Chapter 1). In some cases it will be important to prioritise which remnants will be the most appropriate ones for management. For example, it may not be appropriate to target large remnants in good condition because there is less to improve and hence little to be gained from additional management effort. Conversely, some kinds of degraded remnants like those on highly productive valley floors and in riparian areas may be more valuable to focus on for management interventions. This is because they may improve markedly through actions like fencing, reduced grazing pressure and weed removal. They also may be the places subject to the greatest threats of ongoing degradation. Developing priorities for management will often be best done through a farm plan – a topic we visit in the final chapter of this book (see Chapter 7).

References

1. Cunningham, R.B., Lindenmayer, D.B., Crane, M., Michael, D.R., MacGregor, C., Montague-Drake, R. and Fischer, J. 2008. The combined effects of remnant vegetation and tree planting on farmland birds. *Conservation Biology* **22**: 742–752.

2. Gibbons, P. and Freudenberger, D. 2006. An overview of methods used to assess vegetation condition at the scale of the site. *Ecological Management and Restoration* 7: S10–S7.
3. Gibbons, P., Briggs, S.V., Ayers, D.A., Doyle, S., Seddon, J., McElhinny, C., Jones, N., Sims, R. and Doody, J.S. 2008. Rapidly quantifying reference conditions in modified landscapes. *Biological Conservation* **141**: 2483–2493.
4. Montague-Drake, R., Lindenmayer, D.B. and Cunningham, R.B. 2009. Factors affecting site occupancy by woodland bird species of conservation concern. *Biological Conservation* **142**: 2896–2903.
5. Stagoll, K., Manning, A.D., Knight, E., Fischer, J. and Lindenmayer, D.B. 2010. Using bird-habitat relationships to inform urban planning. *Landscape and Urban Planning* (in press): doi:10.1016/j.landurbplan.2010.1007.1006.
6. Michael, D.R., Cunningham, R.B. and Lindenmayer, D.B. 2008. A forgotten habitat? Granite inselbergs conserve reptile diversity in fragmented agricultural landscapes. *Journal of Applied Ecology* **45**: 1742–1752.
7. Cunningham, R.B., Lindenmayer, D.B., Crane, M., Michael, D. and MacGregor, C. 2007. Reptile and arboreal marsupial response to replanted vegetation in agricultural landscapes. *Ecological Applications* **17**: 609–619.
8. Barton, P.S., Manning, A.D., Gibb, H., Lindenmayer, D.B. and Cunningham, S. 2009. Conserving ground-dwelling beetles in an endangered woodland community: multi-scale habitat effects on assemblage diversity. *Biological Conservation* **142**: 1701–1709.
9. McElhinny, C., Lowson, C., Schneemann, B. and Pachon, C. 2009. Variation in litter under individual tree crowns: Implications for scattered tree ecosystems. *Austral Ecology* **35**: 87–95.
10. Barton, P.S., Manning, A., Gibbs, H., Cunningham, S. and Lindenmayer, D.B. 2010. Fine-scale heterogeneity in beetle assemblages under co-occurring *Eucalyptus* in the same subgenus. *Journal of Biogeography* **37**: 1927–1937.
11. Kirkpatrick, J.B., Gilfedder, L., Bridle, K. and Zacharek, A. 2005. The positive and negative conservation impacts of sheep grazing and other disturbances on the vascular plant species and vegetation of lowland subhumid Tasmania. *Ecological Management and Restoration* **6**: 51–60.
12. Dorrough, J. and Scroggie, M.P. 2008. Plant responses to agricultural intensification. *Journal of Applied Ecology* **45**: 1274–1283.
13. McIntyre, B.S. and Tongway, D. 2005. Grassland structure in native pastures: links to soil surface condition. *Ecological Management and Restoration* **6**: 43–50.
14. Prober, S.M. and Lunt, I. 2008. Kangaroo Grass: a keystone species for restoring weed-invaded temperate grassy woodlands. *Australian Plant Conservation* **17**: 22–23.
15. Grey, M.J. 2008. The impact of the Noisy Miner *Manorina melanocephala* on avian diversity and insect-induced dieback in eucalypt remnants. PhD thesis.

School of Life Sciences, Department of Zoology. La Trobe University: Melbourne.

16. Wilkins, S., Keith, D.A. and Adam, P. 2003. Measuring success: evaluating the restoration of a grassy eucalypt woodland on the Cumberland Plain, Sydney, Australia. *Restoration Ecology* **11**: 489–503.
17. Prober, S.M., Thiele, K.R., Lunt, I.D. and Koen, T.B. 2005. Restoring ecological function in temperate grassy woodlands: manipulating soil nutrients, exotic annuals and native perennial grasses through carbon supplements and spring burns. *Journal of Applied Ecology* **42**: 1073–1085.
18. Munro, N.T., Fischer, J., Wood, J. and Lindenmayer, D.B. 2009. Revegetation in agricultural areas: the development of structural complexity and floristic diversity. *Ecological Applications* **19**: 1197–1210.
19. Dorrough, J., Moxham, C., Turner, V. and Sutter, G. 2006. Soil phosphorus and tree cover modify the effects of livestock grazing on plant species richness in Australian grassy woodland. *Biological Conservation* **130**: 394–405.
20. Briggs, S.V., Taws, N.M., Seddon, J.A. and Vanzella, B. 2008. Condition of fenced and unfenced remnant vegetation in inland catchments in south-eastern Australia. *Australian Journal of Botany* **56**: 590–599.
21. Gibson-Roy, P., Delpratt, J. and Moore, G. 2007. Restoring Western (basalt) plains grassland: 2. Field emergence, establishment and recruitment following direct seeding. *Ecological Management and Restoration* **8**: 123–132.
22. Driscoll, D., Milkovits, G. and Freudenberger, D. 2000. 'Impact and use of firewood in Australia'. CSIRO Sustainable Ecosystems Report. CSIRO: Canberra.
23. Manning, A.D., Lindenmayer, D.B. and Cunningham, R.B. 2007. A study of coarse woody debris volumes in two grassy box-gum woodland reserves in the Australian Capital Territory. *Ecological Management and Restoration* **8**: 221–224.
24. Cenci, F.B., Louvandini, H., McManus, C.M., Dell'Porto, A., Costa, D.M., Araujo, S.C. and Abdulla, A.L. 2007. Effects of condensed tannin from *Acacia mearnsii* on sheep infected naturally with gastrointestinal helminthes. *Veterinary Parasitology* **144**: 132–137.
25. Crane, M., Lindenmayer, D.B. and Cunningham, R.B. 2010. The use of den trees by the squirrel glider (*Petaurus norfolcensis*) in temperate Australian woodlands. *Australian Journal of Zoology* **58**: 39–49.
26. Spooner, P.G. and Briggs, S.V. 2008. Woodlands on farms in southern New South Wales: a longer-term assessment of vegetation changes after fencing. *Ecological Management and Restoration* **9**: 33–41.
27. Fischer, J., Stott, J., Zerger, A., Warren, G., Sherren, K. and Forrester, R.I. 2009. Reversing a tree regeneration crisis in an endangered ecoregion. *Proceedings of the National Academy of Sciences* **106**: 10386–10391.

28. Watson, D.M. 2001. Mistletoe – a keystone resource in forests and woodlands worldwide. *Annual Review of Ecology and Systematics* **32**: 219–249.
29. Crane, M., Montague-Drake, R.M., Cunningham, R.B. and Lindenmayer, D.B. 2008. The characteristics of den trees used by the Squirrel Glider (*Petaurus norfolcensis*) in temperate Australian woodlands. *Wildlife Research* **35**: 663–675.
30. Manning, A.D., Lindenmayer, D.B. and Barry, S.C. 2004. The conservation implications of bird reproduction in the agricultural 'matrix': a case study of the vulnerable superb parrot of south-eastern Australia. *Biological Conservation* **120**: 363–374.
31. Weinberg, A., Gibbons, P., Briggs, S.V. and Bonser, S. 2010. The extent and pattern of *Eucalyptus* regeneration in an agricultural landscape. *Biological Conservation* (in press): doi:10.1016/j.biocon.2010.08.020.
32. Close, D.C., Davidson, N.J. and Watson, T. 2008. Health of remnant woodlands in fragments under distinct grazing regimes. *Biological Conservation* **141**: 2395–2402.
33. Lunt, I.D. 1994. Variation in flower production of nine grassland species with time since fire, and implications for grassland management and restoration. *Pacific Conservation Biology* **1**: 359–366.
34. Prober, S.M.S., Spindler, L.H. and Brown, A.H.D. 1998. Conservation of the grassy white box woodlands: effects of remnant population size on genetic diversity in the allotetraploid herb *Microseris lanceolata*. *Conservation Biology* **12**: 1279–1290.
35. Fischer, J. and Lindenmayer, D.B. 2002. The conservation value of paddock trees for birds in a variegated landscape in southern New South Wales. 2. Paddock trees as stepping stones. *Biodiversity and Conservation* **11**: 833–849.
36. Manning, A.D., Gibbons, P. and Lindenmayer, D.B. 2009. Scattered trees: a complementary strategy for facilitating adaptive responses to climate change in modified landscapes? *Journal of Applied Ecology* **46**: 915–919.
37. Gibbons, P. and Boak, M. 2002. The value of paddock trees for regional conservation in an agricultural landscape. *Ecological Management and Restoration* **3**: 205–210.
38. Grey, M.J., Clarke, M.F. and Loyn, R.H. 1997. Initial changes in the avian communities of remnant eucalypt woodlands following a reduction in the abundance of noisy miners, *Manorina melanocephala*. *Wildlife Research* **24**: 631–648.
39. Grey, M.J., Clarke, M.F. and Loyn, R.H. 1998. Influence of the Noisy Miner *Manorina melanocephala* on avian diversity and abundance in remnant Grey Box woodland. *Pacific Conservation Biology* **4**: 55–69.
40. MacDonald, M.A. and Kirkpatrick, J.B. 2003. Explaining bird species composition and richness in eucalypt-dominated remnants in subhumid Tasmania. *Journal of Biogeography* **30**: 1415–1426.

41. Oldland, J.M., Taylor, R.S. and Clarke, M.F. 2009. Habitat preferences of the Noisy Miner (*Manorina melanocephala*) – a propensity for prime real estate? *Austral Ecology* **34**: 306–316.
42. Lindenmayer, D.B., Knight, E.J., Crane, M.J., Montague-Drake, R., Michael, D.R., and MacGregor, C.I. 2010. What makes an effective restoration planting for woodland birds? *Biological Conservation* **143**: 289–301.
43. Manning, A.D. and Lindenmayer, D.B. 2009. Paddock trees, parrots and agricultural production: an urgent need for large-scale, long-term restoration in south-eastern Australia. *Ecological Management and Restoration* **10**: 126–135.
44. Lindenmayer, D.B., Cunningham, R.B., MacGregor, C., Tribolet, C. and Donnelly, C.F. 2001. A prospective longitudinal study of landscape matrix effects on fauna in woodland remnants: experimental design and baseline data. *Biological Conservation* **101**: 157–169.
45. Lindenmayer, D.B., Cunningham, R.B., MacGregor, C., Crane, M., Michael, D., Fischer, J., Montague-Drake, R., Felton, A. and Manning, A. 2008. Temporal changes in vertebrates during landscape transformation: a large-scale 'natural experiment'. *Ecological Monographs* **78**: 567–590.

3

What makes a good planting?

In a nutshell

A good planting will typically have several or all of the following features:

- comprises native plants
- incorporates small remnants of native vegetation (e.g. patches of large diameter living or dead paddock trees)
- contains a mixture of species of native trees and shrubs (particularly *Acacia*) to provide a range of kinds of habitats for native wildlife
- preferably be block-shaped or at least 40 metres wide to provide some protected interior areas away from the planting's edges
- be located near to areas of remnant native woodland or other plantings
- be managed (e.g. with appropriate fencing) to either exclude or control livestock access to limit the amount of grazing and trampling pressure
- be managed to control pest plants and animals.

Introduction

Replanting native vegetation is a major land management practice throughout extensive parts of Australia's agricultural zone. It can have many ecological and other environmental benefits. Moreover, recent financial incentive schemes for biodiversity conservation on farms now explicitly recognise the importance of planted areas of native vegetation on farms. Some incentive schemes are more

Box 3.1. Burnbank – a planting success story extraordinaire

The property of Burnbank is near Ladysmith on the South West Slopes of New South Wales. It is owned and managed by Rick and Pam Martin. When the Martins bought their farm in the late 1970s, there was about 2% tree cover and it was beset by major problems like salinity and rising water tables. They embarked on an ambitious revegetation program and plantings now cover more than 14% of the farm with several areas of large block-shaped plantings (see Figure 3.1).

Problems with salinity and rising water tables at Burnbank have been solved. There has also been an extraordinary response by wildlife. Detailed surveys by researchers from The Australian National University over the past nine years have identified many native birds of conservation significance in the plantings at Burnbank. These include the Speckled Warbler, Diamond Firetail, Southern Whiteface, Red-capped Robin, Flame Robin, Hooded Robin and Crested Shrike-tit.

There have been many other successes associated with restoration efforts in Australia's agricultural areas. For example, well-targeted revegetation efforts in north-eastern Victoria have been instrumental in assisting the recovery of the Grey-crowned Babbler.

likely to provide funds to those farmers willing to establish plantings that benefit biodiversity and other key ecological values.

Our work over the past decade[1–3] as well as research by other scientists (e.g. [4, 5]) has shown that how a planting is done can significantly influence how effective it is for biodiversity conservation. Indeed, when plantings are done well, they can be true success stories, highlighting effective ways to integrate conservation and production objectives on a well-managed farm.

In this chapter, we discuss the characteristics of good plantings for wildlife. We cover such key topics as where you should plant, the size of plantings, the shape and other aspects of the geometry of plantings, and what you should plant. In essence, this material covers the appropriate context, configuration and content of plantings. Of course, the effectiveness of plantings can be strongly influenced by how they are managed after they have been established. On this basis, the final section of this chapter examines some management issues for plantings.

While it may not be possible to implement all of the things we recommend in this chapter on every farm, for the vast majority of farms, some areas of planting will be far better than having none. There is a good reason for this. We have found that plantings are quite different habitats for wildlife than areas of remnant native vegetation[6, 7] (Box 3.2). Therefore, a farm with plantings **and** remnant native vegetation will typically support more species of native animals (e.g. birds) than a farm with no plantings. Plantings are therefore an important part of the 'portfolio of vegetation assets' for biodiversity on a farm.[8]

Figure 3.1: (a) Large block planting. (Photo by David Lindenmayer). (b) Rick and Pam Martin, managers of Burnbank Farm near Wagga Wagga, and (c) large area of revegetated land on Burnbank Farm. (Photos by Sachiko Okada)

Box 3.2. Planting-specialist bird species

A detailed study of the response of birds to different kinds of vegetation on a farm has shown that some birds are plantings specialists. That is, they are significantly more likely to be found in plantings than any other sort of vegetation on a farm. They include several species thought to be declining and that are of conservation concern, such as the Rufous Whistler, Red-capped Robin, Scarlet Robin and Speckled Warbler.[7]

The structure of the vegetation within plantings is generally more dense and the ground layer has higher levels of cover than elsewhere on a farm. These features appear to provide suitable feeding and nesting sites for these planting-specialist bird species.[3]

Having some plantings on a farm is better than having no plantings

As in Box 3.2, we have used birds to illustrate many of the key topics in this chapter. There are several good reasons for this, but the primary one is that birds have received far greater attention from researchers than reptiles, mammals, invertebrates, plants or indeed any other group. A second reason (partially related to the first) is that many species of birds seem to respond quickly and positively to plantings.[1, 3] This includes many species of birds that breed successfully within plantings, sometimes only a few years after they are established. Despite the bias toward birds in this chapter, we have included examples from other groups wherever possible.

Where should you plant?

Near streams and watercourses

The most species-rich plantings for birds are those established around watercourses

Many landholders and natural resource managers ask if there are places where plantings can be located to maximise their effectiveness for wildlife. Work that we have recently completed has indicated that bird species diversity is highest in plantings established in and around watercourses and riparian areas, particularly where the land is flat.[3] On average, there are more than three additional species of birds in these areas than in plantings away from watercourses on steep slopes. Riparian areas provide ready access to open water for drinking, and areas around streams and watercourses can support higher levels of nutrients and water flow than elsewhere in a landscape. This promotes growth of trees and other plants which can be important determinants of habitat suitability for some animals.

Figure 3.2: (a) Densely stocked planting. (Photo by Nicola Munro). (b) Revegetated farm. (Photo by Nicola Munro). (c) Red-capped Robin, a bird that responds strongly to planting. (Photo by Julian Robinson)

Box 3.3. Management objectives for plantings

There can be different reasons why a landholder might wish to establish revegetated areas on a farm. These may include establishing a wind break, providing shade for livestock, limiting soil erosion, tackling salinity, enhancing property value, improving aesthetics, firewood production (or some other kind of farm forestry activity), or improving conservation values. Key questions for guiding any management activity are: What are your objectives? What do you want to achieve from your management intervention? These are important questions to pose **before** embarking on any project. This is because objectives can (and usually will) significantly influence what you do and the way you do it. For example, the size, structure and species composition of a planting will be different if it is primarily for farm forestry, as opposed to creating shelter for livestock.

Our focus in this book is on improving conservation outcomes, and the way a planting is established can make a big difference to how suitable it is as habitat for particular species as well as the overall diversity and abundance of plants and animals. Indeed, you might do a planting quite differently if an objective is to maximise the number of bird species (i.e. species richness), as opposed to creating habitat for a particular species of bird such as a threatened or endangered one.[3] For example, if the aim is to encourage species such as the Glossy Black Cockatoo then a high proportion of the planting would be made up of Drooping She-oak. If the aim is to establish areas suitable for the Squirrel Glider, then plantings might be located in and around existing remnants such as paddock trees.

In some cases it will be possible to derive multiple kinds of environmental benefits from the same action. For example, establishing a wind break planting is not only valuable for improved stock production but it also can create useful habitat for some species of native animals. A narrow wind break planting will be different in its biodiversity values than a wide wind break planting, however, with many species more likely to breed successfully in wider plantings.

Plantings established near other plantings or areas of remnant native vegetation will support the most species of birds

Plantings in gully lines appear to be used more frequently for feeding by species like the charismatic but threatened Squirrel Glider. In some cases, however, plantings in gully lines need careful management to address issues like weed invasion. This is because plantings in gully lines are more likely to support larger numbers of exotic weed species than plantings elsewhere in the landscape.[9]

Near other plantings and areas of remnant native vegetation

An emerging principle from an array of studies is that many species of animals respond strongly to the overall amount of habitat in a landscape. Plantings established near other plantings or close to areas of existing remnant native

Figure 3.3: A planting in a riparian area. (Photo by Nicola Munro)

woodland support more species of birds and are more likely to support a range of individual bird species like the Grey Fantail and the Rufous Whistler.[3] The reasons for these findings may be that the neighbourhood of a planting which supports

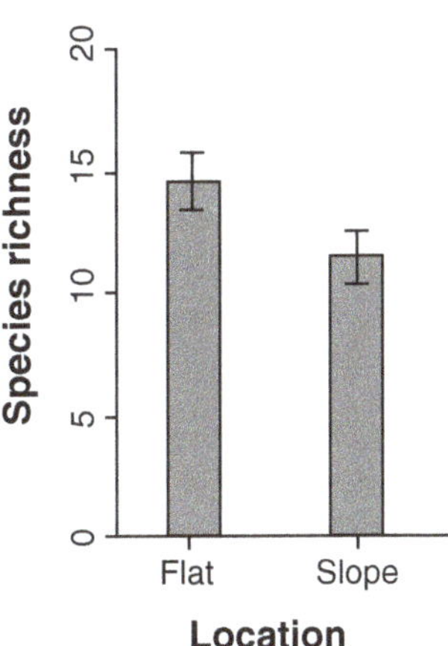

Figure 3.4: Bird species richness in flat areas around streams and watercourses and on steep slopes away from watercourses.

Figure 3.5: Birds that are often found in farm plantings: (a) The Rufous Whistler, a species of conservation concern. (Photo by Julian Robinson). (b) The Willie Wagtail. (Photo by Damian Michael)

remnant vegetation or planted native vegetation may provide additional habitat for species and/or facilitate the movement of organisms between habitat patches.

Scattered paddock trees are a critical type of remnant native vegetation on a farm[10–12] (see Chapter 4), and we have found that plantings established around paddock trees can significantly increase their suitability for bird species like the White-browed Woodswallow and the White-plumed Honeyeater.[3] The combination of large old remnant paddock trees and adjacent young, densely planted trees may increase the range of places where birds can find food or construct a nest. The value of paddock trees as nodal points around which to establish plantings needs to be weighed against the potential for natural regeneration to occur around paddock trees if processes like the intensity of

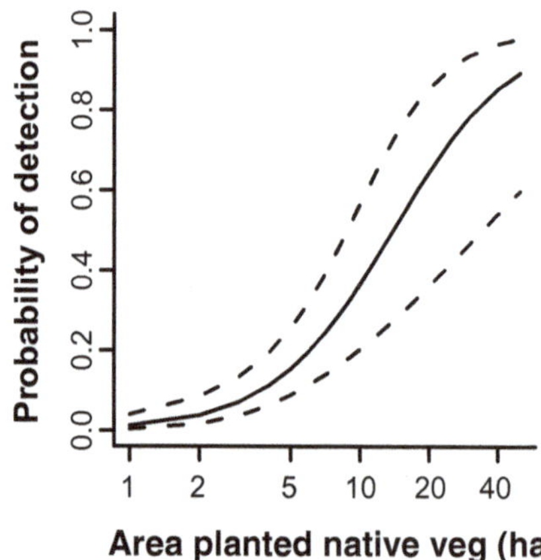

Figure 3.6: Relationships between the occurrence of the Rufous Whistler in a planting and the amount of surrounding planted vegetation.

Figure 3.7: Planted areas that also support scattered paddock trees. (Photos by Nicola Munro)

Figure 3.8: White-browed Woodswallow. Several species of woodswallows are listed as birds of conservation concern because they are thought to be declining. Of these species, the White-browed Woodswallow is one that often uses replanted native vegetation. (Photo by Julian Robinson)

Paddock trees and cleared areas with many logs can be valuable places around which to establish plantings

grazing pressure are relieved – even for a relatively short period. This is an important assessment for land managers to make because natural regeneration is cheaper than planting, but it requires different kinds of grazing regimes than traditional set stocking.[13]

Like scattered paddock trees, fallen timber is also now known to be a critically important habitat feature used by many species of native animals (see below) ranging from invertebrates to birds.[3, 14, 15] Given this, cleared areas characterised by large amounts of fallen timber may be valuable ones to target for planting, particularly as recent work has demonstrated strong positive relationships between bird species richness (i.e. the number of different bird species present) in plantings and the prevalence of fallen timber.[3]

Some places not to plant

Just as there are some key areas where plantings are likely to be most effective, there are also some parts of farms best to avoid when establishing plantings. For example, intact areas of natural native grasslands are very important habitats and it is inappropriate to damage, degrade or even totally destroy them by targeting them for planting. Many areas of open grazing country such as on the South West Slopes of New South Wales are native secondary grasslands, however, where in most cases plantings would be beneficial if established appropriately. Similarly, rocky areas can be critically important environments for species-rich groups of native reptiles[16] (see Chapter 5) and planting programs in these areas need to be carefully considered. Densely spaced plantings in rocky areas can be particularly detrimental because of changes to the amount of incoming solar radiation and, in turn, thermal conditions for reptiles.[17]

How big should your planting be?

The general scientific wisdom is that large plantings of two or more hectares are better for biodiversity than small ones (e.g. those less than 1 hectare). This is because they are thought likely to support more species of plants and animals than small ones and support larger numbers of individuals of a given species. We have found that some species are indeed more likely to occur in larger plantings. An example is the White-plumed Honeyeater.[3] The odds of finding this species in a planting increase by 27% with each doubling of the area of a planting. Other species which are more likely to occur in larger plantings include the Apostlebird and the Red-browed Finch.[7]

Big plantings are often better than small plantings, but in many cases the context of a planting is even more important. The amount of native vegetation or planted vegetation surrounding a planting has a very strong effect on the occurrence of animals such as birds.

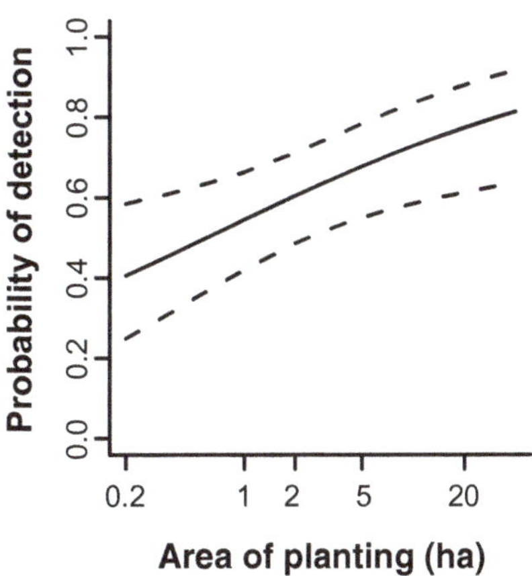

Figure 3.9: Relationships between the size of a planting and the occurrence of the White-plumed Honeyeater.

Figure 3.10: White-plumed Honeyeater feeding a Pallid Cuckoo chick. Cuckoos lay their eggs in the nests of other birds, a phenomenon known as brood parasitism. (Photo by Julian Robinson)

Although large plantings tend to be better for biodiversity than small ones, it is not always possible to establish large plantings on a farm. In these cases, establishing small plantings is still better than having no areas of revegetation on a farm.

What shape should a planting be?

There are two broad shapes of plantings – narrow strips or linear-shaped plantings, and block plantings. The former are typically 20–40 metres wide and have often been established between paddocks. Block plantings are wider plantings that are typically square or rectangular in shape. In our experience these are often established on hilltops or in the corners of large paddocks.

Several field studies have indicated that block plantings support more species and higher populations of individual species of birds than narrow linear plantings or strip plantings.[2] For example, the Rufous Songlark is more likely to occur in wider block plantings than narrow plantings that are less than 30 metres wide. Other species more likely to occur in block-shaped plantings include the Red-capped Robin and the Apostlebird.[7] A likely reason for planting shape effects is that some birds breed more successfully in wide plantings than in narrow ones. This may be because rates of predation on nests are lower in wide strips.[18]

Figure 3.11: Block planting. (Photo by David Lindenmayer)

Bird species richness in isolated strip plantings is generally lower than block plantings. Some birds, such as the Red Wattlebird and Grey Shrike-thrush, are most likely to be found in block plantings. Linking isolated strip plantings to create

Figure 3.12: The Rufous Songlark prefers wider block plantings. (Photo by Julian Robinson)

Box 3.4. How wide should a planting be?

Many landholders have asked us how wide plantings should be to make them valuable for farm wildlife. As we have outlined in this chapter, any plantings on a farm, even narrow ones (under 30 metres wide) are better than no plantings. Recent work, however, is suggesting that some species of birds prefer to nest at least 17 metres away from the edges of a planting. Therefore, a 40-metre wide planting would provide a band of just six metres of interior area within a planting (see Figure 3.13). This indicates that plantings 40–60 metres or more in width will be those where nesting is most likely to be successful.

We are aware that many plantings are narrower than 40–60 metres. This does not mean they are without value, even though there may be some species which will not nest successfully within them. Moreover, there may be opportunities to widen these planted areas with additional rows of adjacent trees and shrubs or to link narrow plantings with other plantings or patches of remnant native vegetation.[2]

intersections (see Figure 3.14) can significantly increase the number of species of birds they contain. This means where there are space constraints on a farm that might preclude the establishment of large block plantings, an alternative approach

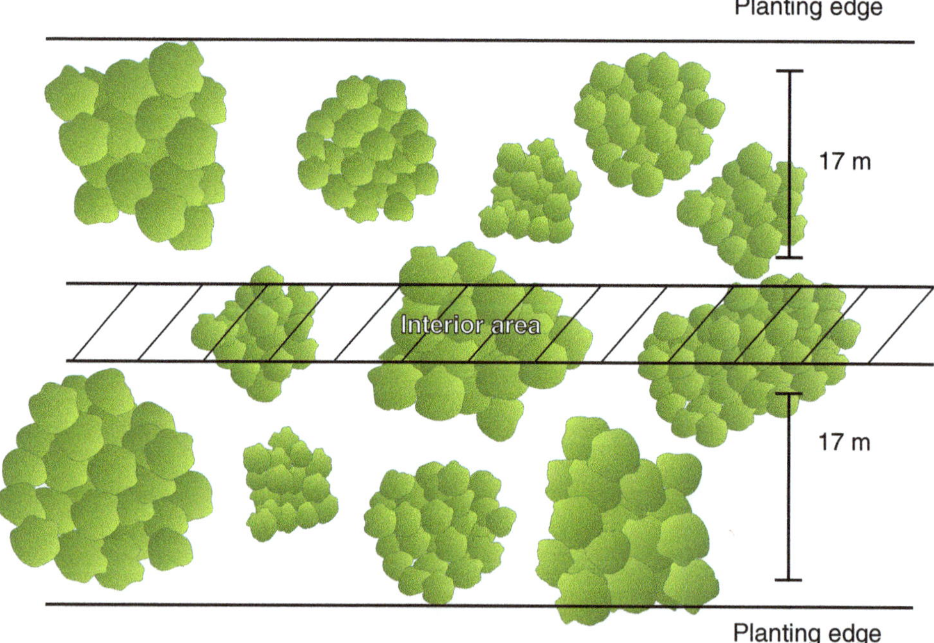

Figure 3.13: The configuration of edge and interior areas for bird nesting success in a planting. (Drawing by Clive Hilliker)

Figure 3.14: Intersecting plantings. (Photo by David Lindenmayer)

with positive outcomes for native birds might be to establish linked sets of intersecting strip plantings around paddock perimeters.[2]

What should you plant?

Plantings should always be established with native trees and shrubs

Much has been written in the past about what should be planted in revegetation programs. An increasing body of scientific research is clearly showing that plantings are best established with locally native species of trees and shrubs rather than exotic species. There are several key reasons for this:

1. Native shrubs established within planted areas reduce the time needed for some species of native birds to colonise these areas.[19]
2. Plantings established with native plants are less likely to be used by exotic bird species like the House Sparrow, Common Starling and European Blackbird. Conversely, native bird species richness is lower in plantings with more exotic species.[19]
3. Plantings established with particular species of native trees like the Drooping She-oak can provide valuable habitat for high-profile threatened vertebrates like the Glossy Black Cockatoo.

Figure 3.15: Glossy Black Cockatoo – a bird species which responds positively to plantings with mature species like Drooping She-oak. (Photo by Eleanor Sobey)

4. Local native trees support more species of native invertebrates than exotic trees.[20] In turn, invertebrates are important sources of food for many native birds and mammals.
5. Exotic species may colonise areas well beyond planted areas and eventually become environmental weeds elsewhere in agricultural landscapes. Such environmental weeds can be expensive to remove and in some cases virtually impossible to control.
6. Plantings established with native trees are more resilient to fire. The vast majority of native plants have well-developed abilities to recover naturally after fire, particularly if they are well established (e.g. greater than 10–15 years old) before they are burned. This is in marked contrast to most exotic species of plants.
7. Incentive payments to establish plantings may not (and should not) be forthcoming if revegetated areas are dominated by exotic or invasive plant species.
8. Locally native plants are more likely to regenerate naturally which means a tree planting will be there for the long term.

9. Many exotic species such as tree lucerne are so invasive that they can choke an entire tree planting which, over time, can result in the planting being dominated by a single species.

Establishing suitable plant species goes beyond simply selecting those species that are generally local to a given area. There is also a need to consider suitable microclimatic conditions for particular plants. For example, different species of trees, understorey shrubs and ground cover plants grow better on ridges and rocky slopes than on valley floors. Establishing tree and understorey species in appropriate places can influence not only rates of growth, but also rates of mortality and ultimately the success of a revegetation program.

The understorey layer of plantings is a critical vegetation layer for native biota. For example, understorey cover is an important part of the suitability of plantings for native bird species such as the Grey Fantail, Rufous Whistler and the White-plumed Honeyeater[3] as well as the use of these areas for feeding by the Squirrel Glider. Plantings with a dense midstorey wattle layer often do not support the hyper-aggressive Noisy Miner – a native honeyeater that is notorious for driving away many small native woodland birds[21] (see Box 2.7 in Chapter 2). Establishing an understorey in plantings is therefore important, otherwise revegetation programs risk creating more habitat for the Noisy Miner with detrimental additional impacts on other native birds.[22]

The value of the understorey layer in plantings highlights the value for biodiversity of using mixtures of tree and shrubs in revegetated areas.[9, 19] Indeed, this is a general principle not only for planted vegetation in temperate woodland environments[1] but also replanted areas in Australian tropical forest landscapes (e.g. [23]).

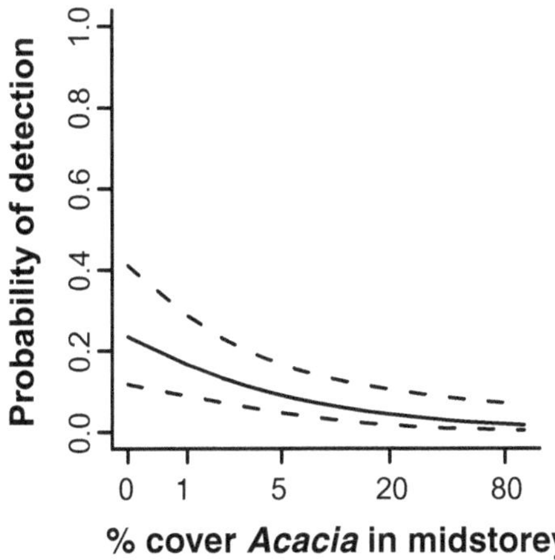

Figure 3.16: Relationships between the occurrence of the Noisy Miner and the density of wattle trees in the understorey of a planting.

Box 3.5. Birds versus reptiles – dealing with the complexity created by different responses of different animals groups to plantings

The previous paragraph highlighted how the density of the understorey in a planting can have a positive effect on bird species richness. Dense plantings are not good for all species, however. For example, rock-dwelling and ground-dwelling reptiles can be strongly disadvantaged by dense vegetation cover because it limits the amount of light (and hence heat) reaching the ground.[17] This highlights two common themes in conservation science and natural resource management: (1) the need to think deeply about the objectives of management; for example, is the aim to establish a planting for birds or for reptiles? (see Box 1.2); and (2) one prescription does not fit all. If an aim of management is to create suitable environments for a range of different species on a farm, then the best way to deal with the complexity created by different species is to make sure that not all plantings on a farm are established in exactly the same way. For example, some plantings might have densely stocked understorey vegetation to make them unsuitable for the Noisy Miner, whereas others might have trees and other plants spaced widely to provide suitable thermal micro-environmental conditions for some species of reptiles.

Management of plantings

The way plantings are managed can have a substantial influence on their suitability for wildlife. For example, plantings are often characterised by lower levels of grazing pressure than other parts of farms and this can significantly influence their use by some species, such as the Superb Fairy-wren[3] (Figure 3.17) and the Olive Legless Lizard (Box 3.6). Total exclusion of grazing from plantings

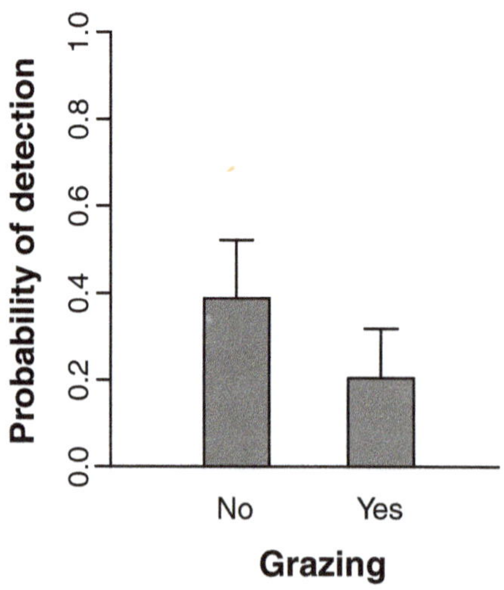

Figure 3.17: Relationships between the occurrence of the Superb Fairy-wren in plantings and grazing.

Box 3.6. Plantings as a haven for geckos without legs

The Olive Legless Lizard superficially resembles a snake. It differs from snakes, however, by having distinct ear openings. It is, in many respects, a gecko without legs. This is because legless lizards and geckoes are very closely related and share many characteristics such as a fleshy tongue and squeaky vocalisations.

The Olive Legless Lizard is one of the species that we have found to be abundant in plantings[6], particularly under artificial substrates like railway sleepers that we have placed in these areas. The habitat requirements of this species often include grassy tussocks where it can forage. The regulation of grazing pressure may make fenced planted areas more suitable than other parts of farms for the Olive Legless Lizard.

may not be feasible on some farms, particularly during prolonged periods of drought when other sources of food for livestock are limited. Plantings, however, should not be subject to intense and prolonged set stocking grazing. Rather, grazing regimes like low-intensity and/or short-term grazing may be appropriate. Another key part of managing grazing in plantings is to exclude stock until

Limited grazing pressure in plantings is one of the reasons why they are important environments for many species

Figure 3.18: Livestock grazing in a planting. (Photo by David Lindenmayer)

Figure 3.19: Olive Legless Lizard. (Photo by Damian Michael)

trees and shrubs are sufficiently well established to be able to withstand being badly damaged by livestock.

Fencing is one of the usual ways to either exclude grazing or control grazing pressure in plantings. Well-constructed fences may last for 15–20 years – about the same amount of time since many plantings in agricultural areas were first established. Maintaining fences or replacing them with new ones is therefore critical to maintaining the suitability of planted areas for wildlife. Where possible, it is important not to use barbed wire on the top strand of a fence around a planting. Many kinds of animals ranging from bats to marsupial gliders and birds can become entangled in barbed wire and die (see Chapter 2). For those fences where it is not feasible to exchange barbed wire for straight wire, it can be useful to enclose barbed wire in a poly-pipe covering or tie flagging tape to the barbed strand to increase visibility.

Dead trees, shrubs and fallen logs are critical habitat for many kinds of native animals and they should not be removed from plantings. For example, dead trees are often used as nesting sites by the Superb Parrot and the Squirrel Glider.[12, 24, 25] Similarly, dead shrubs in plantings form an important part of the habitat of the White-browed Woodswallow. Bird species richness is highest in plantings where fallen branches and large logs are retained. Logs can play many key ecological roles for biodiversity in plantings, ranging from providing places for birds to perch and call (e.g. Rufous Songlark), to facilitating animal foraging through providing habitat for invertebrate prey.[3, 15]

Maintain or replace fences around plantings to limit the risks of overgrazing and trampling

Box 3.7. The virtual fence

Researchers at CSIRO have applied technology from the new millennium to the humble farm fence. By fitting cattle with collars containing a global positioning system (GPS) and a device that can deliver an irritating sound, they have created, in effect, a virtual fence. Cattle that stray close to the 'fence' are automatically bombarded with the annoying sound until they retreat to the desired area. If they hit the 'fence' they receive an electric shock from the collar that is similar to a conventional electric fence. Researchers found that cattle could learn the whereabouts of the 'fence' in less than an hour. The technology enables a farmer to track stock via the internet and alter the configuration of fences either manually or automatically. This could be a valuable tool for regulating stock movements in remnant vegetation or plantings. While the system is still at a prototype stage, it may become cost-effective as the cost of the technology decreases relative to the cost of fencing.

Given the importance of fallen timber for many animals, there can be considerable ecological value in adding old fence posts or even branch trimmings to the ground layer of plantings. Similarly, the absence of trees with hollows from most plantings means that nest boxes are of value in revegetated areas to promote the colonisation of some cavity-dependent animals like gliders and bats.[26, 27]

Keep dead shrubs and trees in plantings – don't remove them

Plantings can sometimes become a refuge for unwanted pest species on farms. For example, they can harbour populations of the Red Fox, European Rabbit and an array of weeds. Activities like ripping rabbit warrens (but see Box 5.3 in Chapter 5) and laying poison baits for feral predators are therefore an important part of the management of plantings. Plantings in

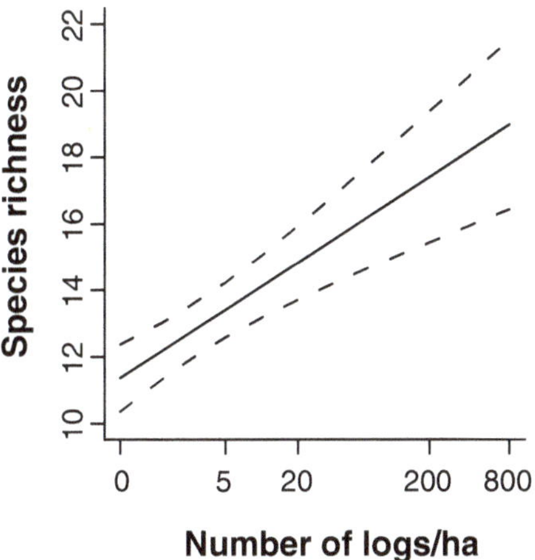

Figure 3.20: Relationships between the prevalence of logs in plantings and bird species richness.

Figure 3.21: Adding fallen timber to Mulligan's Flat Nature Reserve in Canberra. Dead eucalypts were gathered from areas cleared for suburban development and then distributed throughout the reserve to create habitat for animals such as reptiles and invertebrates. (Photo by Adrian Manning)

Undertake regular pest animal and weed control in plantings

riparian areas have a particular propensity to become weedy over time and weed control is important in these areas.[9]

Summary

Plantings are significant environments on farms and they provide valuable habitat for a wide range of species, although their importance has been best demonstrated for birds.[1] Plantings should be considered as part of the portfolio of vegetation assets on a farm because they support a different suite of species than other kinds

Box 3.8. Fires and plantings

Some landowners are concerned that additional plantings on a farm will make their property more fire-prone. Fire proneness and fire risk reduction are complex topics that are beyond the scope of this book. We note that plantings can significantly reduce wind speeds on a farm, however, and in turn can reduce the speed at which fires can burn through crops and paddocks. A key issue is that access points through plantings may need to be planned in advance to ensure that a landowner has an escape route to the leeward side of a planting in the unfortunate event of a wildfire. During the Junee fires in 2006, firefighters found that dense linear tree plantings along the perimeter of paddocks sometimes created barriers, cutting off escape routes. This is easily overcome by leaving a few gaps in the plantings so firefighters can simply cut the fence to escape the oncoming fire.

Figure 3.22: Burned planted area during the 2006 Junee wildfires. (Photo by Will Barton)

Figure 3.23: Common Ringtail Possum – a species often found on some farms but rarely in replanted native vegetation. (Photo by Mike Greer)

Box 3.9. Are areas of native vegetation and plantings interchangeable?

Much of our work over the past decade has indicated that areas of remnant native temperate woodland and replanted areas are not interchangeable environments. On average, areas of remnant native vegetation support three times more bird species than plantings.[7] Therefore it is not possible to clear an area of remnant native woodland from one part of a farm and then attempt to offset its loss by establishing a planting elsewhere on that farm, without having a significant net negative effect on biodiversity.[28] We have found that for an average farm, while the establishment of plantings will increase overall bird diversity by two extra species, clearing native vegetation will reduce bird diversity by at least seven species and many of these will be species of conservation concern. Species at risk of being lost from farms where native vegetation is cleared include the Hooded Robin and Jacky Winter.[7]

A key reason why remnant native temperate woodland and replanted areas are not interchangeable environments is because they provide distinctly different kinds of habitat for native animals; that is, these two broad kinds of vegetation are characterised by quite different sets of animals and plants.[6, 7, 9, 19] The Common Ringtail Possum (see Figure 3.23), the Common Brushtail Possum and the Brown Treecreeper are examples of species rarely found in plantings. This is because the trees in plantings are typically not old enough to provide critical nesting resources like hollow-bearing trees. Similarly, plantings lack large fallen trees that are part of the habitat of the Brown Treecreeper. This further underscores why: (a) it will take a long time before the features which characterise patches of old-growth woodland will eventually develop in replanted vegetation, and (b) it is important to protect existing areas of native vegetation.

In summary, it has become increasingly clear that while both patches of remnant native vegetation and plantings are valuable habitats on farms, it is inappropriate to substitute one for the other.[28] Rather, it is far better to consider them as complementary habitats which, when considered in combination, will contribute significantly greater diversity and numbers of animals on a farm than will either one in isolation.[7]

of native vegetation, such as old-growth woodland. Such complementarity means that it is not ecologically appropriate to substitute plantings for cleared remnant native vegetation on a farm (Box 3.9).

The effectiveness of plantings for wildlife conservation can be maximised through careful consideration of two key factors when they are being established. These are where plantings are established on a farm and what kinds of trees and understorey plants are used to establish a planting. Once plantings are established, the way they are managed also can have a substantial effect on their habitat suitability for wildlife. Management activities to maintain the suitability of

Box 3.10. When to plant and when to let natural regeneration occur

Establishing new areas of plantings can be an important part of farm management but in some cases it also can be expensive and time consuming. It is therefore important to carefully weigh up when it is appropriate to instigate a planting program and when re-establishing vegetation cover is done best by letting natural regeneration occur. The most successful natural tree regeneration will occur in those areas of a farm that are:

- within approximately two tree heights of living trees, **and**
- subject to limited grazing pressure from domestic livestock, **and**
- where there has been an absence or limited past history of fertiliser application.

Restoration of tree cover in other parts of the farm where there is high grazing pressure or a prolonged history of fertiliser application might be best achieved through an active program of tree planting.[29, 30]

plantings for wildlife include: (1) limiting grazing and trampling pressure by domestic livestock; (2) erecting and then maintaining appropriate fencing; and (3) not removing dead trees, dead shrubs and fallen branches and logs.

References

1. Munro, N., Lindenmayer, D.B. and Fischer, J. 2007. Faunal response to revegetation in agricultural areas of Australia: a review. *Ecological Management and Restoration* **8**: 199–207.
2. Lindenmayer, D.B., Cunningham, R., Crane, M., Michael, D. and Montague-Drake, R. 2007. Farmland bird responses to intersecting replanted areas. *Landscape Ecology* **22**: 1555–1562.
3. Lindenmayer, D.B., Knight, E.J., Crane, M.J., Montague-Drake, R., Michael, D.R. and MacGregor, C.I. 2010. What makes an effective restoration planting for woodland birds? *Biological Conservation* **143**: 289–301.
4. Martin, W.K., Eyears-Chaddock, M., Wilson, B.R. and Lemon, J. 2004. The value of habitat reconstruction to birds at Gunnedah, New South Wales. *Emu* **104**: 177–189.
5. Vesk, P. and Mac Nally, R. 2006. The clock is ticking: revegetation and habitat for birds and arboreal mammals in rural landscapes of southern Australia. *Agriculture, Ecosystems and Environment* **112**: 356–366.
6. Cunningham, R.B., Lindenmayer, D.B., Crane, M., Michael, D. and MacGregor, C. 2007. Reptile and arboreal marsupial response to replanted vegetation in agricultural landscapes. *Ecological Applications* **17**: 609–619.

7. Cunningham, R.B., Lindenmayer, D.B., Crane, M., Michael, D.R., MacGregor, C., Montague-Drake, R. and Fischer, J. 2008. The combined effects of remnant vegetation and tree planting on farmland birds. *Conservation Biology* **22**: 742–752.
8. Lindenmayer, D.B. and Likens, G.E. 2010. *Effective Ecological Monitoring.* CSIRO Publishing: Melbourne.
9. Munro, N.T., Fischer, J., Wood, J. and Lindenmayer, D.B. 2009. Revegetation in agricultural areas: the development of structural complexity and floristic diversity. *Ecological Applications* **19**: 1197–1210.
10. Manning, A.D., Lindenmayer, D.B. and Barry, S.C. 2004. The conservation implications of bird reproduction in the agricultural 'matrix': a case study of the vulnerable superb parrot of south-eastern Australia. *Biological Conservation* **120**: 363–374.
11. Manning, A.D., Fischer, J. and Lindenmayer, D.B. 2006. Scattered trees are keystone structures – implications for conservation. *Biological Conservation* **132**: 311–321.
12. Crane, M., Lindenmayer, D.B. and Cunningham, R.B. 2010. The use of den trees by the squirrel glider (*Petaurus norfolcensis*) in temperate Australian woodlands. *Australian Journal of Zoology* **58**: 39–49.
13. Fischer, J., Stott, J., Zerger, A., Warren, G., Sherren, K. and Forrester, R.I. 2009. Reversing a tree regeneration crisis in an endangered ecoregion. *Proceedings of the National Academy of Sciences* **106**: 10386–10391.
14. Driscoll, D., Milkovits, G. and Freudenberger, D. 2000. 'Impact and use of firewood in Australia'. CSIRO Sustainable Ecosystems Report. CSIRO: Canberra.
15. Barton, P.S., Manning, A.D., Gibb, H., Lindenmayer, D.B. and Cunningham, S.A. 2009. Conserving ground-dwelling beetles in an endangered woodland community: multi-scale habitat effects on assemblage diversity. *Biological Conservation* **142**: 1701–1709.
16. Michael, D.R., Cunningham, R. and Lindenmayer, D.B. 2008. A forgotten habitat? Granite inselbergs conserve reptile diversity in fragmented agricultural landscapes. *Journal of Applied Ecology* **45**: 1742–1752.
17. Michael, D.R., Lindenmayer, D.B. and Cunningham, R.B. 2010. Managing rock outcrops to improve biodiversity conservation in Australian agricultural landscapes. *Ecological Management and Restoration* **11**: 43–50.
18. Major, R.E., Christie, F.J., Gowing, G. and Ivison, T.J. 1999. Elevated rates of predation on artificial nests in linear strips of habitat. *Journal of Field Ornithology* **70**: 351–364.
19. Munro, N., Fischer, J., Barrett, G., Wood, J., Leavesley, A. and Lindenmayer, D.B. 2010. Bird's response to revegetation of different structure and floristics

– are 'restoration plantings' restoring bird communities? *Restoration Ecology* (in press): doi:10.1111/j.1526-1100X.2010.00703.x.

20. Majer, J.D., Recher, H.F., Graham, R. and Watson, A. 2001. 'The potential of revegetation programs to encourage invertebrates and insectivorous birds'. Curtin University School of Environmental Biology Bulletin No. 21. Curtin University: Perth.
21. Grey, M.J., Clarke, M.F. and Loyn, R.H. 1997. Initial changes in the avian communities of remnant eucalypt woodlands following a reduction in the abundance of noisy miners, *Manorina melanocephala*. *Wildlife Research* **24**: 631–648.
22. Maron, M. 2007. Threshold effect of eucalypt density on an aggressive avian competitor. *Biological Conservation* **136**: 100–107.
23. Kanowski, J., Catterall, C.P. and Wardell-Johnson, G.W. 2005. Consequences of broadscale timber plantations for biodiversity in cleared forest landscapes of tropical and subtropical Australia. *Forest Ecology and Management* **208**: 359–372.
24. Manning, A.D., Lindenmayer, D.B. and Cunningham, R.B. 2007. A study of coarse woody debris volumes in two grassy box-gum woodland reserves in the Australian Capital Territory. *Ecological Management and Restoration* **8**: 221–224.
25. Crane, M., Montague-Drake, R.M., Cunningham, R.B. and Lindenmayer, D.B. 2008. The characteristics of den trees used by the Squirrel Glider (*Petaurus norfolcensis*) in temperate Australian woodlands. *Wildlife Research* **35**: 663–675.
26. Suckling, G.C. and Macfarlane, M.A. 1983. Introduction of the Sugar Glider, *Petaurus breviceps*, into re-established forest in the Tower Hill State Game Reserve, Victoria. *Australian Wildlife Research* **10**: 249–258.
27. Smith, G.C. and Agnew, G. 2002. The value of 'bat boxes' for attracting hollow-dependent fauna to farm forestry plantations in southeast Queensland. *Ecological Management and Restoration* **3**: 37–46.
28. Gibbons, P. and Lindenmayer, D.B. 2007. Offsets for land clearing: no net loss or the tail wagging the dog? *Environmental Management and Restoration* **8**: 26–31.
29. Dorrough, J. and Moxham, C. 2005. Eucalypt establishment in agricultural landscapes and implications for landscape-scale restoration. *Biological Conservation* **123**: 55–66.
30. Manning, A.D. and Lindenmayer, D.B. 2009. Paddock trees, parrots and agricultural production: an urgent need for large-scale, long-term restoration in south-eastern Australia. *Ecological Management and Restoration* **10**: 126–135.

4

What makes a good paddock for biodiversity?

In a nutshell

A good paddock will typically have several or all of the following features:

- some areas of scattered trees that are in good condition
- some young regenerating trees to replace older paddock trees as they age and die
- trees of a range of ages, including dead trees
- some areas of fallen timber and native pasture
- appropriate grazing regimes that limit the risks of overgrazing and excessive trampling in high traffic areas by domestic livestock.

Introduction

Paddocks are where grazing and/or cropping take place and these commodity production areas usually dominate a farm. Paddocks have traditionally been thought to be places lacking habitat value for wildlife and unsuitable for biodiversity. Recent research, however, is demonstrating that this is not the case. This work shows that the way paddocks are managed can make a significant difference to the conservation of biodiversity on a farm.

In this chapter, we examine the attributes that can make a paddock good for biodiversity without reducing productivity. We also outline the processes which can degrade the suitability of paddocks for biodiversity and suggest management approaches to address these threatening processes. We dedicate an entire chapter to a discussion of paddocks because they collectively comprise a large proportion

Box 4.1. Biodiversity and paddock productivity – the success story of the dung beetle

The development of large numbers of sheep and cattle in Australia brought with it a problem – large quantities of dung. The dung provided favourable conditions for large numbers of pest flies including those responsible for fly strike in sheep. Concentrations of dung also were thought to contribute to dieback in trees and in some cases reduced the area suitable for livestock grazing within paddocks. Australian native dung beetles (Figure 4.1) are adapted to woodlands, not open pastures, and are better suited to using the fibrous pellets produced by native marsupials.[5] Although native dung beetles can occasionally be found in cattle and horse dung, they are not effective at breaking down the large volumes of dung produced by domestic livestock. Between 1968 and 1982, the CSIRO imported 55 species of dung beetles, mostly from South Africa.[6] Many of these beetles successfully established around Australia, with 17 known to have established in New South Wales.[5] Dung beetles can improve pastures by breaking down dung and burying it underground for their larvae to eat when they hatch. This has the dual effect of cycling nutrients from the dung into the soil, as well as removing dung from the surface and reducing the area of otherwise fouled grazing land. An additional benefit is the reduction of pest fly populations that lay their eggs in dung. Introduced dung beetles complement our native dung beetles and are a good example of the link between biodiversity and ecological function, with the end result being improved paddock productivity.

of production landscapes and therefore represent an important opportunity for biodiversity conservation. Government and non-government organisations in many countries around the world, including Australia, are increasingly recognising this. They are establishing programs that provide payments to farmers to manage paddocks in ways that are sympathetic to the needs of biodiversity (e.g. [1–4]).

Figure 4.1: Native dung beetle (*Onthophagus* sp.). (Photo by Philip Barton)

Paddock trees

Paddock trees are keystone structures in agricultural landscapes and play many valuable ecological roles for farm wildlife

Scattered paddock trees are legacies of the previous, often more dense, cover of native woodland. Most of the tree cover in many agricultural areas comprises paddock trees and these trees typically pre-date European settlement. Our studies have demonstrated that paddock trees have many critical values for woodland biodiversity[7–11] and we list some of them in Box 4.2. In fact, paddock trees can be so valuable in agricultural landscapes that they have been termed 'keystone structures'. This means that paddock trees have an array of roles which makes them disproportionately valuable relative to the area they occupy.[9, 11] Paddock trees often have characteristics that are rare or uncommon in younger trees on a farm. For example, paddock trees can have: (1) large cavities or hollows in the trunk or large lateral branches; (2) a large and/or deep canopy; (3) prolific numbers of flowers; (4) many clumps of mistletoe; and (5) large quantities of flaking bark. Large branches and a deep canopy are, in turn, important sources of fallen timber and leaf litter which provide habitat for a range of animals.[12, 13] The environment immediately around paddock trees also can be very important for native plants.

After they have died, paddock trees can remain important nesting and denning sites for wildlife, including for high profile species like the Superb Parrot[14] and the

Figure 4.2: A scattered paddock tree landscape near Holbrook in southern New South Wales. (Photo by David Lindenmayer)

Figure 4.3: The Varied Sittella – a native bird which depends on large branches that occur on very large trees. (Photo by Suzi Bond)

Squirrel Glider.[11] Of course, dead standing trees eventually collapse, but as we discuss in the following section, fallen timber is also a critically important kind of habitat for many native plants and animals on farms.

Threats to paddock trees and threat management

The urgency of the problem

There is a range of threats to the ongoing persistence of good populations of scattered paddock trees in Australian rural landscapes. There is limited, if any, regeneration of paddock trees in many areas.[25] This, combined with land clearing and the high rate of mortality among paddock trees from spray drift, stock camps and ageing of existing old trees (see Box 4.3), means there is a high risk of large areas being devoid of paddock trees within the next century. This major problem needs to be addressed urgently. This is because it takes several hundred years for paddock trees to develop to a size where they provide certain ecological functions or habitat resources; for example, tree hollows. Therefore, a delay in tree

Box 4.2. Biodiversity values of scattered paddock trees for woodland ecosystems

Scattered paddock trees have many important values for wildlife on farms. They:

- contribute to the range of kinds of vegetation cover on farms. This, in turn, is a highly significant factor influencing bird species richness on farms[15]
- increase the suitability of adjacent woodland remnants for declining woodland birds (e.g. the Brown Treecreeper, Jacky Winter, Black-chinned Honeyeater)[16]
- provide habitat for a very high diversity of woodland invertebrates[17]
- act as stepping stones for the movement of birds through agricultural landscapes[7, 18]
- provide habitat for a range of species of reptiles such as the Marbled Gecko and Carnaby's Wall Skink[19]
- provide key nesting trees for the Superb Parrot[14]
- provide nesting trees and feeding trees for the Squirrel Glider[11, 20]
- provide nesting and roosting trees for bats[21]
- provide places around which young trees can regenerate naturally[10, 22]
- provide sites around which to establish plantings[23] (see Chapter 3)
- increase the suitability of adjacent plantings for the White-browed Woodswallow and the White-plumed Honeyeater[23] (see Chapter 3)
- raise overall bird species richness, as well as the richness of woodland bird specialists, to levels comparable to denser woodland patches.[24]

Figure 4.4: A large paddock tree can provide many types of shelter and food resources for wildlife. (Photo by Damian Michael)

Figure 4.5: Red Wattlebird – a species that uses paddock trees as stepping stones to move through agricultural landscapes. (Photo by Rachel Muntz)

Figure 4.6: Superb Parrot – a species that often nests in paddock trees. (Photo by Julian Robinson)

regeneration now will prolong the time that landscapes support few or no mature paddock trees. Finding alternatives to clearing paddock trees is important because this remains one of the greatest threats to these trees in rural areas.[22, 26]

Limiting grazing pressure to perpetuate paddock trees

Well-managed grazing can reduce direct and indirect damage that stock can cause to paddock trees. Direct damage includes stock harming a tree's bark or eating its leaves. Indirect damage includes fatally high levels of nutrients from concentrated levels of animal dung entering the soil. Large amounts of dung also may lead to large increases in the numbers of insects which can defoliate trees (see Box 4.3).

Scattered paddock trees can be perpetuated in landscapes by reducing grazing pressure through cell or rotational grazing practices to enable regeneration to occur.[10] Fencing areas around groups of paddock trees also can prevent livestock from camping under them.

Planning cropping to better protect paddock trees

Paddock trees are being cleared in some regions because they are incompatible with modern farming practices, particularly the use of agricultural infrastructure such as centre pivots, lateral-moving irrigation systems,[27] and large GPS-guided machinery for precision agriculture such as tram-lining.

One possible way to integrate cropping with the long-term maintenance of populations of scattered paddock trees is to employ a zoning system where parts of paddocks are designed as places where existing paddock trees will be retained and new ones will be recruited through strategies that promote tree regeneration.

Limiting spray drift and fertiliser use to promote paddock tree protection

Paddock trees can be particularly sensitive to herbicide spray drift. Several farm management activities can help reduce the susceptibility of paddock trees to dieback (see Box 4.3). One of these is to regulate the application of chemical sprays carefully. Another strategy that reduces mortality among paddock trees is avoiding the build-up of nutrients around trees from fertiliser. Reducing the amount of nutrients from fertiliser application around paddock trees can promote the natural regeneration of overstorey woodland trees.

Limiting firewood harvesting

The firewood industry is a major threat to populations of scattered paddock trees.[28] Individual farm practices that regulate or prohibit commercial firewood harvesting are a vital management tool that can significantly contribute to the preservation of scattered standing and fallen paddock trees.

Figure 4.7: Micro-fencing around paddock trees to better protect them. (Photo by David Lindenmayer)

Control the impacts of fire on paddock trees

Large old trees can be particularly fire-prone and can be badly damaged or even killed outright by a high severity burn. A good way to protect paddock trees is to restrict the use of fire around them.

Complementary plantings around paddock trees

Reduced grazing pressure is critical to ensure that regeneration of paddock trees can take place

Paddock trees can be valuable places around which to establish block or strip plantings (see Chapter 3). Plantings with paddock trees have proved to be more valuable habitats for some native animal species than plantings without paddock trees.[23, 29]

Fallen timber

Fallen timber and large logs are increasingly recognised for the wide range of important ecological roles they play on farms. These include:

- storing large amounts of carbon
- storing large quantities of nutrients which are eventually returned to the soil as a log decomposes
- trapping leaf litter and soil

Box 4.3. What causes dieback in paddock trees?

Since the mid to late 1960s, the premature decline and death of trees in rural landscapes has increased markedly. This phenomenon is known as rural dieback. Dieback has been reported in all States in Australia and affects a wide range of forest and woodland types. The problem has become particularly pronounced in some areas. For example, Platt[30] reported that approximately 28% of the trees in a 3300-hectare area of pastoral land in north-eastern Victoria died in a 22-year period between 1971 and 1993. If this trend continues, then almost all remaining trees in the area will be dead in about 80 years. Dieback is also becoming increasingly common in south-eastern Queensland.[31]

Dieback is most serious where less than 30% of native vegetation cover remains[31] but there is no single cause of dieback in paddock trees.[32] Some of the contributing factors include ([33–35]):

- defoliation by native invertebrates (such as stick insects) and introduced insects
- over-abundance of the aggressive native honeyeater, the Noisy Miner
- application of nutrients to improve pastures, which can lead to changes in soil fertility, and changes in insect grazing pressure
- impacts of plant pathogens such as fungi
- salinity
- soil acidification
- drought
- fire
- deterioration of the structure of the soil (e.g. soil compaction)
- mechanical damage caused by livestock and farm machinery
- increased numbers of parasitic mistletoes
- airborne salt
- decreased abundance of wildlife, such as the Sugar Glider, which can consume large quantities of tree-defoliating insects, and the Echidna, which eats insect larvae.

- increased infiltration and retention of water in the soil
- providing places for plants and fungi to germinate
- providing foraging, breeding and sheltering sites for many kinds of animals including invertebrates, frogs, reptiles, native mammals and birds
- creating basking sites for reptiles and perching places for birds
- acting as micro-fire breaks to slow the spread of fire
- providing runways to facilitate native animal movement through landscapes
- providing shelter sites for livestock, particularly lambs.

Many of our studies over the past decade have clearly demonstrated the value of logs and fallen timber for biodiversity. As an example, adding fallen timber to paddocks significantly increases populations of native mammals like the Fat-tailed

Figure 4.8: Eucalypt dieback in southern New South Wales. (Photo by Nicola Munro)

Dunnart. A study examining wildlife use of recycled wooden fence posts distributed throughout the native grasslands of Terrick Terrick National Park found the Fat-tailed Dunnart constructed nests beneath the fence posts within a month. The number of Dunnarts recorded beneath the posts increased over a five-month period, peaking during the spring breeding season.[36] Fallen timber in paddocks also adds significantly to the range and quality of habitats for birds on farms. Overall bird species richness on a farm is higher on those farms where paddocks support fallen timber.[15]

In another of our research projects, we have found that logs in paddocks support highly diverse assemblages of native beetles, including many natural predators of pest insects (Figure 4.11). These log-associated beetle assemblages have proved to be markedly different from those found in open paddocks or under trees.[17] This work highlights the extraordinary diversity of invertebrates that can characterise Australian farm and woodland environments. The importance of fallen timber for beetles is illustrated by the number of species which are added for

Figure 4.9: Fat-tailed Dunnart. (Photo by Damian Michael)

every additional log within a paddock. Figure 4.11 summarises the number of beetle species identified from a survey of 30 logs and patches of open ground in a grassy eucalypt woodland. It shows how most of the beetle species from areas of open ground can be found after examining only 10 different patches. In contrast,

Figure 4.10: Predatory ground beetles shelter under dead wood. (Photo by Philip Barton)

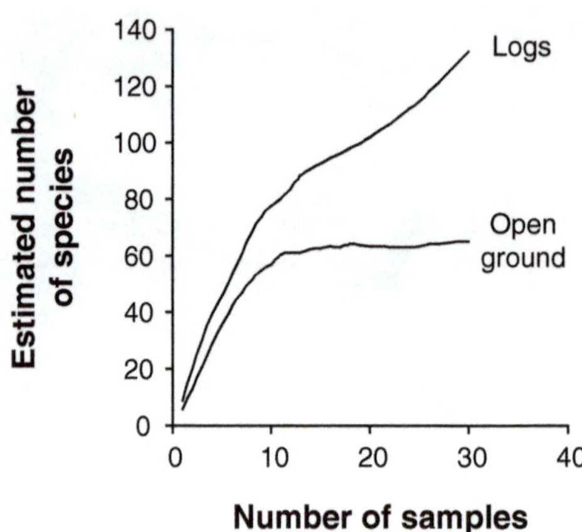

Figure 4.11: The number of beetle species from open ground and logs in a grassy eucalypt woodland.

the number of beetle species found around logs continues to climb, with twice as many species found at 30 logs compared with 30 open patches.

Threats to fallen timber and their management

Firewood removal can cause a significant loss and degradation of fallen timber resources (see Box 4.5). Therefore, careful management to limit the extent of firewood collection will have positive benefits for farm wildlife. Preventing firewood contractors from removing fallen timber and cutting down dead standing trees is the best way to halt the depletion of log resources on a farm. There is also a need for ongoing lobbying to encourage urban Australians to use alternative kinds of home heating to firewood. Cool climate cities like Melbourne, Canberra, Albury–Wodonga and Wagga Wagga import massive quantities of firewood from threatened woodland environments.

Box 4.4. A falling fenceline but rising biodiversity

Old fence posts left on the ground to decompose provide important refuges for many different kinds of animals.[37] In the grasslands of Terrick Terrick National Park, a two-kilometre fallen fenceline consisting of 270 posts provided habitat for the threatened Curl Snake. Over a nine-month period, 38 Curl Snakes were observed sheltering beneath the fence posts. These small (40–60 cm) nocturnal snakes were not the only animals to use fallen fence posts. Geckos, legless lizards, skinks and frogs all make use of log refuges.[36] Many of these species spend almost their entire lives sheltering beneath the same log, or group of logs, and will often forage within a short distance from their home-site. Old fence posts left on the gound alongside newly established fences provide small reptiles and native mammals with a habitat corridor.

Figure 4.12: Curl Snake. (Photo by Damian Michael)

Box 4.5. Firewood and biodiversity

Considerable debate has been associated with native forest logging and the effects of plantation forestry in Australia. There is a third industry, however, that has largely escaped public attention – the firewood industry. More than 4.5 million tonnes of firewood are cut annually for domestic consumption in Australia.[28, 38] In 2003, this was approximately two-thirds of the volume of woodchips exported each year. The 2001 State of the Environment Report[39] noted that firewood is the third largest source of energy in Australia after gas and coal.

The firewood industry is based on cutting living and dead standing and fallen trees, particularly on private land and on roadside reserves. These trees and the vegetation from which they are removed can have significant value for many elements of biodiversity – including some threatened ones such as the Superb Parrot.[40] Much of the timber is sourced from temperate woodlands that have been heavily cleared and are subject to other threatening processes, such as overgrazing by domestic livestock, rural dieback and salinity.[41] Indeed, in New South Wales, firewood cutting is listed as a key threatening process in temperate woodlands and is having an enormous impact on the many species strongly associated with dead standing trees and fallen timber.[28] For example, Garnett and Crowley identified 21 species of birds at risk from firewood harvesting.[42]

Figure 4.13: Firewood cutting in a patch of temperate woodland on the South West Slopes of New South Wales. (Photo by Esther Beaton)

Relocating fallen timber

Fallen timber can be a problem for machinery in cropping paddocks. In these cases, a management solution may be to collect the timber and redistribute it within woodland remnants or in plantings where it can play a valuable role as part of the habitat for a range of species, including threatened or declining native woodland birds[16, 23] (see Chapters 2 and 3). Because of the importance of logs within plantings, existing paddocks with large amounts of fallen timber can be good ones to target for revegetation.

Commercial firewood cutting is a highly destructive industry that has many negative damaging effectives on farm wildlife. The activities of commercial firewood contractors should be limited or excluded on a farm

Fallen timber is obviously derived from standing trees. Therefore, maintaining an ongoing long-term supply of fallen timber in a paddock will require the retention of paddock trees as well as the recruitment of new trees that will eventually replace existing ones.

Native grasses

There is a long tradition in Australia of establishing many kinds of exotic pasture grasses for domestic livestock grazing, and areas formerly dominated by temperate

Figure 4.14: Swards of native grasses. Native grasslands in temperate agricultural areas are now one of the most endangered ecological communities in Australia. (Photos by (a) Damian Michael and (b) David Lindenmayer)

Figure 4.15: Tessellated Gecko – a species of reptile often associated with native grasses. (Photo by Damian Michael)

woodlands are no different in this regard. It is increasingly clear, however, that exotic ground cover can have significant negative effects on plant and animal diversity. For example, species like the Eastern Yellow Robin are less likely to occur in areas dominated by exotic grasses. Conversely, woodland remnants surrounded by paddocks with predominantly native grass cover are significantly more likely to be occupied by bird species like the Diamond Firetail.[16] Native grass cover also has an important positive effect on reptile diversity[43] and is one of the kinds of vegetation assets on a farm that contributes significantly to overall bird diversity at the property level.[15]

Threats to native grasses and their management

The ecological integrity of native pastures can be negatively affected by overgrazing as well as some of the practices typically linked with traditional grazing regimes like the application of chemical fertilisers. One of the best ways to restore native pastures is to alter grazing regimes and adopt more environmentally sensitive ones like rotational grazing or cell grazing. Native grasses and other kinds of native plants often found in native pastures (e.g. forbs) are sensitive to the high levels of nitrogen created through repeated fertiliser application. One

Having some areas of native grass or native pasture on a farm can provide significant benefits for farm wildlife. Native pastures are also likely to survive droughts better and can recover quickly after wildfire

Figure 4.16: Native grasslands provide habitat for the endangered Golden Sun Moth, which needs Wallaby Grass to complete its life cycle. Invasion of weeds and overgrazing by sheep and cattle threaten this endangered insect. (Photo by Suzi Bond)

way this problem has been addressed has been to apply carbon to pastures (in the form of sugar).[46] This may be feasible over only relatively small areas, however, and might not be financially or logistically practical over extensive paddocks.

Because native pastures can sometimes be overlooked, they can be threatened by inappropriate tree establishment programs such as plantations of exotic tree species (e.g. Radiata Pine). Plantings of native woodland trees have occasionally

Box 4.6. The economic benefits of native grasses

It is increasingly recognised that native grasses can have economic benefits for the bottom line on a farm as well as substantial ecological benefits. Many species of native grasses have comparable levels of protein to some traditional exotic pasture species. Native grasses, particularly perennials, are better adapted to acidic and saline soils than pastures dominated by exotic grasses, as well as being more drought-tolerant and persistent under adverse seasonal conditions.[44] For example, on the South West Slopes of New South Wales, few exotic pasture species are able to take advantage of summer rain. In contrast, native grasses can respond rapidly after just a few millimetres of rain, and provide quality feed for livestock at a time when it would otherwise be scarce. Finally, the establishment of swards of native grasses (e.g. Kangaroo Grass) can help control exotic weed species.[45]

Figure 4.17: (a) The Archers' farm near Gundagai, New South Wales, where considerable effort has been expended in developing pastures dominated by native grasses that have helped reduce the effects of drought on their farm and livestock. (Photo by Sam Archer). (b) Sam and Sabrina Archer. (Photo courtesy of Sam Archer)

also resulted in the degradation or destruction of native grassland. Careful assessment of areas proposed for plantation development or revegetation programs should help avoid these problems.

Shrubs and saltbush

A range of native shrubs can provide important stock fodder. While the palatability of many such shrubs is often low relative to grass, many shrubs provide an important source of nutrition when other forage is scarce. Good examples include members of the Chenopod or saltbush family, which are particularly high in protein. While Lucerne has a crude protein value of 16.3%, Old Man Saltbush can have a crude protein value of 21.9%.[47] Table 4.1 provides examples of some species and their relative palatability.

As well as being an important food resource for stock, the presence of shrubs in a paddock greatly enhances its habitat value by providing both food and shelter for native wildlife.

The shelter value of some kinds of shrubs is substantial and they offer many small animals, particularly a range of reptiles and small birds, protection from predators. Shrubs also can provide inconspicuous nesting sites for many birds, particularly ground-foraging birds like fairy-wrens. The scattering of shrubs across an otherwise open paddock is valuable for reptiles, allowing them to thermoregulate effectively (going in and out of the shade as they get too hot or cool). Kangaroos may also shelter under low-growing shrubs, making 'hip holes'. These hip holes are important germination sites for many native plants and eventually become microhabitats for other animals, such as lizards and small mammals.[49]

Shrubs are also important sources of food. For instance, Ruby Saltbush has red berries that are consumed by many species of birds and reptiles, such as the Shingleback. The fruit of the Dillon Bush (said to taste like salty grapes) is also eaten by many birds and indeed the germination of this plant is aided by passing through the gut of emus.[48] The succulent foliage of many of the saltbushes (such as the copperburrs and bluebushes) is important for lizards, such as the Bearded Dragon,[50] while kangaroos may occasionally forage on flat-leaved saltbushes (*Atriplex* spp.).[51] Seeds from a range of chenopods (including *Atriplex*, *Maireana*, *Chenopodium* and *Sclerolaena* spp.) are important food sources for the Plains Wanderer.[52] Creeping Saltbush and Climbing Saltbush not only have fruit that is eaten by native birds, but are an important food source for caterpillars of native butterflies and moths, which in turn attract insect-eating birds. Lignum is rich in pollen and nectar and is therefore another important food source for native animals.[47]

More about saltbush

The research team from The Australian National University has recently commenced a study to examine which species of birds and reptiles occupy Old

Table 4.1: Examples of native shrubs offering complementary stock fodder

Common name	Scientific name	Palatability*
Old Man Saltbush	*Atriplex nummularia*	Eaten readily when other forage is scarce
Bladder Saltbush	*Atriplex vesicaria*	Eaten after more palatable plants have been removed
Creeping Saltbush	*Atriplex semibaccata*	Although not keenly sought by stock, it is perhaps the most readily grazed of the saltbushes
Nitre Goosefoot	*Chenopodium nitrariaceum*	Utilised during stress periods, considered a valuable drought reserve
Cannon-ball	*Dissocarpus paradoxus*	Utilised when palatable annual forbs have been grazed
Climbing Saltbush	*Einadia nutans*	Useful forage plant
Ruby Saltbush	*Enchylaena tomentosa*	Provides a useful reserve for drought periods
Cottonbush	*Maireana aphylla*	Utilised in stress periods
Yanga Bush	*Maireana brevifolia*	Not browsed while other forage is available but withstands browsing moderately well
Black Bluebush	*Maireana pyramidata*	Generally regarded as drought forage only
Thorny Saltbush	*Rhagodia spinescens*	Moderately palatable, browsed heavily when other forage is scarce
Short-winged Copperburr	*Sclerolaena brachyptera*	Relatively palatable, reliable forage species
Grey Copperburr	*Sclerolaena diacantha*	Heavily utilised when more acceptable forbs and grasses are not available, has high nutritive value
Woolly Copperburr	*Sclerolaena lanicuspis*	Considered good forage, particularly in early growth stages
Lignum**	*Muehlenbeckia florulenta*	May be used when other feed is scarce
Dillon Bush**	*Nitraria billardieri*	Leaves may be browsed by sheep when feed is otherwise scarce

* Palatability notes from Brooke and McGarva.[48]

** Not members of the Chenopod family.

Man Saltbush plantings. Our results are preliminary, but the animals already found to inhabit saltbush plantings include 31 species of birds (including the Red-capped Robin and Red-chested Button-quail), one species of native small mammal (the Fat-tailed Dunnart), one species of snake (the Curl Snake) and six species of lizards. Nonethless, such monoculture plantings probably offer less habitat value than a more natural system comprised of scattered saltbushes interspersed with species of native grasses.

Many species of saltbush are susceptible to overgrazing and trampling which impairs their root systems and hence the capacity to extract limited soil moisture

Figure 4.18: Saltbush planting. (Photo by Mason Crane)

during dry periods. Overgrazing is most likely to occur during dry times (most livestock preferring grass if it is available), and hence careful attention must be given to stocking levels during drought. Reducing stock numbers, or using short-term 'crash' grazing, is required to maintain the presence and health of these shrubs during extended dry periods.

Summary

Much of the focus of conservation efforts on farms has been on patches of remnant native vegetation. While this is vitally important, well-managed paddocks also can be a significant part of conservation efforts, particularly if they include one of more of the following features – scattered paddock trees, shrubs, fallen timber, and native pastures. Management practices in paddocks that promote the maintenance or recruitment of these features can include prevention of firewood removal, the establishment of swards of native grasses and areas of native pasture, the maintenance or planting of saltbush, and the maintenance of scattered paddock trees, including the regeneration of new cohorts of trees to replaces ones that are ageing and dying.

References

1. Department of the Environment, Water, Heritage and the Arts. 2009. 'Environmental Stewardship Program. Box-Gum Grassy Woodland Project.' Department of the Environment, Water, Heritage and the Arts: Canberra.
2. Kleijn, D. and Sutherland, W.J. 2003. How effective are European agri-environmental schemes in conserving and promoting biodiversity? *Journal of Applied Ecology* **40**: 947–969.
3. Benayas, J.M.R., Newton, A.C., Diaz, A. and Bullock, J.M. 2009. Enhancement of biodiversity and ecosystem services by ecological restoration: a meta-analysis. *Science* **325**: 1112–1124.
4. Greening Australia. 2009. Introducing Whole of Paddock Rehabilitiation (WOPR): a new approach to regenerating the farm. Available from: http://www.greeningaustralia.org.au/uploads//Our%20Resources%20-%20pdfs/ACT_WOPR09.pdf.
5. Tyndale-Biscoe, M. 1990. *Common Dung Beetles in Pastures of South-Eastern Australia.* CSIRO Division of Entomology: Canberra.
6. Nichols, E., Spector, S., Louzada, J., Larsen, T., Amequita, S. and Favila, M.E. 2008. Ecological functions and ecosystem services provided by Scarabaeinae dung beetles. *Biological Conservation* **141**: 1461–1474.
7. Fischer, J. and Lindenmayer, D.B. 2002. The conservation value of paddock trees for birds in a variegated landscape in southern New South Wales. 2. Paddock trees as stepping stones. *Biodiversity and Conservation* **11**: 833–849.
8. Gibbons, P. and Boak, M. 2002. The value of paddock trees for regional conservation in an agricultural landscape. *Ecological Management & Restoration* **3**: 205–210.
9. Manning, A.D., Fischer, J. and Lindenmayer, D.B. 2006. Scattered trees are keystone structures – implications for conservation. *Biological Conservation* **132**: 311–321.
10. Fischer, J., Stott, J., Zerger, A., Warren, G., Sherren, K. and Forrester, R.I. 2009. Reversing a tree regeneration crisis in an endangered ecoregion. *Proceedings of the National Academy of Sciences* **106**: 10386–10391.
11. Crane, M., Lindenmayer, D.B. and Cunningham, R.B. 2010. The use of den trees by the squirrel glider (*Petaurus norfolcensis*) in temperate Australian woodlands. *Australian Journal of Zoology* **58**: 39–49.
12. Killey, P., McElhinny, C., Rayner, I. and Wood, J. 2010. Modelling fallen branch volumes in a temperate eucalypt woodland: implications for large senescent trees and benchmark loads of coarse woody debris. *Austral Ecology* (in press): doi:10.1111/j.1442-9993.2010.02107.x.
13. McElhinny, C., Lowson, C., Schneemann, B. and Pachon, C. 2010. Variation in litter under individual tree crowns: implications for scattered tree ecosystems. *Austral Ecology* **35**: 87–95.

14. Manning, A.D., Lindenmayer, D.B., Nix, H.A. and Barry, S. 2005. A bioclimatic analysis of the highly mobile Superb Parrot of south-eastern Australia. *Emu* **105**: 193–201.
15. Cunningham, R.B., Lindenmayer, D.B., Crane, M., Michael, D.R., MacGregor, C., Montague-Drake, R. and Fischer, J. 2008. The combined effects of remnant vegetation and tree planting on farmland birds. *Conservation Biology* **22**: 742–752.
16. Montague-Drake, R., Lindenmayer, D.B. and Cunningham, R.B. 2009. Factors affecting site occupancy by woodland bird species of conservation concern. *Biological Conservation* **142**: 2896–2903.
17. Barton, P.S., Manning, A.D., Gibb, H., Lindenmayer, D.B. and Cunningham, S.A. 2009. Conserving ground-dwelling beetles in an endangered woodland community: multi-scale habitat effects on assemblage diversity. *Biological Conservation* **142**: 1701–1709.
18. Fischer, J. and Lindenmayer, D.B. 2002. Small patches can be valuable for biodiversity conservation: two case studies on birds in southeastern Australia. *Biological Conservation* **106**: 129–136.
19. Lindenmayer, D.B., Cunningham, R.B., MacGregor, C., Tribolet, C. and Donnelly, C.F. 2001. A prospective longitudinal study of landscape matrix effects on fauna in woodland remnants: experimental design and baseline data. *Biological Conservation* **101**: 157–169.
20. Crane, M., Montague-Drake, R.M., Cunningham, R.B. and Lindenmayer, D.B. 2008. The characteristics of den trees used by the Squirrel Glider (*Petaurus norfolcensis*) in temperate Australian woodlands. *Wildlife Research* **35**: 663–675.
21. Lumsden, L.F., Bennett, A.F., Silins, J. and Krasna, S. 1994. 'Fauna in a remnant vegetation-farmland mosaic: movements, roosts and foraging ecology of bats'. A report to the Australian Nature Conservation Agency 'Save the Bush' Program. Flora and Fauna Branch, Department of Conservation and Natural Resources: Melbourne.
22. Gibbons, P., Lindenmayer, D.B., Fischer, J., Manning, A.D., Weinberg, A., Sedden, J., Ryan, P. and Barrett, G. 2008. The future of scattered trees in agricultural landscapes. *Conservation Biology* **22**: 1309–1319.
23. Lindenmayer, D.B., Knight, E.J., Crane, M.J., Montague-Drake, R., Michael, D.R. and MacGregor, C.I. 2010. What makes an effective restoration planting for woodland birds? *Biological Conservation* **143**: 289–301.
24. Stagoll, K., Manning, A.D., Knight, E., Fischer, J. and Lindenmayer, D.B. 2010. Using bird-habitat relationships to inform urban planning. *Landscape and Urban Planning* (in press): doi:10.1016/j.landurbplan.2010.1007.1006.
25. Weinberg, A., Gibbons, P., Briggs, S.V. and Bonser, S. 2010. The extent and pattern of *Eucalyptus* regeneration in an agricultural landscape. *Biological Conservation* (in press): doi:10.1016/j.biocon.2010.08.020.

26. Gibbons, P., Briggs, S.V., Ayers, D.A., Seddon, J.A., Doyle, S.J., Cosier, P., McElhinny, C., Pelly, V. and Roberts, K. 2009. An operational method to assess impacts of land clearing on terrestrial biodiversity. *Ecological Indicators* **9**: 26–40.
27. Maron, M. and Fitzsimons, J.A. 2007. Agricultural intensification and loss of matrix habitat over 23 years in the West Wimmera, south-eastern Australia. *Biological Conservation* **135**: 587–593.
28. Driscoll, D., Milkovits, G. and Freudenberger, D. 2000. 'Impact and use of firewood in Australia'. CSIRO Sustainable Ecosystems Report. CSIRO: Canberra.
29. Bond, S. 2004. Do woodland birds breed in revegetated sites? Hons thesis. The Australian National University: Canberra.
30. Platt, S. 1999. 'Dieback lessons: learning how to manage sustainably'. Land for Wildlife Notes No. 34, http://www.lowecol.com.au/lfw/gfwmeminfo/Dieback_mgmt_LfW_Note_Vic.pdf.
31. McIntyre, S., McIvor, J.G. and Heard, K.M., eds. 2002. *Managing and Conserving Grassy Woodlands*. CSIRO Publishing: Melbourne.
32. Landsberg, J. and Wylie, F.R. 1983. Water stress, leaf nutrients and defoliation: a model of dieback of rural eucalypts. *Australian Journal of Ecology* **8**: 27–41.
33. Landsberg, J. and Wylie, F.R. 1991. A review of rural dieback in Australia. In *Growback '91*. (Eds Offor, T. and Watson, R.J.) pp. 3–11. Growback Publications: Melbourne.
34. Reid, N. and Landsberg, J. 2000. Tree decline in agricultural landscapes. In *Temperate Eucalypt Woodlands in Australia: Biology, Conservation, Management and Restoration*. (Eds Hobbs, R.J. and Yates, C.J.) pp. 127–166. Surrey Beatty and Sons: Chipping Norton.
35. MacDonald, M.A. and Kirkpatrick, J.B. 2003. Explaining bird species composition and richness in eucalypt-dominated remnants in subhumid Tasmania. *Journal of Biogeography* **30**: 1415–1426.
36. Michael, D., Lunt, I.D. and Robinson, W.A. 2004. Enhancing fauna habitat in grazed native grasslands and woodlands: use of artificially placed log refuges by fauna. *Wildlife Research* **31**: 65–71.
37. Michael, D., Lunt, I.D. and Robinson, W.A. 2003. Terrestrial vertebrate fauna of grasslands and grassy woodlands in the Terrick Terrick National Park, northern Victoria. *Victorian Naturalist* **120**: 164–171.
38. Wall, J. 2000. Fuelwood in Australia: impacts and opportunities. In *Temperate Eucalypt Woodlands in Australia: Biology, Conservation, Management and Restoration*. (Eds Hobbs, R.J. and Yates, C.J.) pp. 372–381. Surrey Beatty and Sons: Chipping Norton.
39. Commonwealth of Australia. 2001. 'State of the Environment Report'. Commonwealth of Australia: Canberra.

40. Manning, A.D., Gibbons, P. and Lindenmayer, D.B. 2009. Scattered trees: a complementary strategy for facilitating adaptive responses to climate change in modified landscapes? *Journal of Applied Ecology* **46**: 915–919.
41. Hobbs, R.J. and Yates, C.J., eds. 2000. *Temperate Eucalypt Woodlands in Australia: Biology, Conservation, Management and Restoration.* Surrey Beatty and Sons: Chipping Norton.
42. Garnett, S.T. and Crowley, G.M. 2000. 'The Action Plan for Australian Birds 2000'. Natural Heritage Trust: Canberra.
43. Cunningham, R.B., Lindenmayer, D.B., Crane, M., Michael, D. and MacGregor, C. 2007. Reptile and arboreal marsupial response to replanted vegetation in agricultural landscapes. *Ecological Applications* **17**: 609–619.
44. McIntyre, S. 2008. The role of plant leaf attributes in linking land use to ecosystem function in temperate grassy vegetation. *Agriculture, Ecosystems and Environment* **128**: 251–258.
45. Prober, S.M. and Lunt, I.D. 2009. Restoration of *Themeda australis* swards suppresses soil nitrate and enhances ecological resistance to invasion by exotic annuals. *Biological Invasions* **11**: 171–181.
46. Prober, S.M., Thiele, K.R., Lunt, I.D. and Koen, T.B. 2005. Restoring ecological function in temperate grassy woodlands: manipulating soil nutrients, exotic annuals and native perennial grasses through carbon supplements and spring burns. *Journal of Applied Ecology* **42**: 1073–1085.
47. Kent, K., Earl, G., Mullins, B., Lunt, I. and Webster, R. 2002. Native Vegetation Guide for the Riverina. Notes for land managers on its management and revegetation. Johnson Centre, Charles Sturt University: Albury, New South Wales.
48. Brooke, G. and McGarva, L. 1998. *The Glove Box Guide to Plants of the NSW Rangelands.* NSW Department of Agriculture: Orange, New South Wales.
49. Eldridge, D.J. and Rath, D. 2002. Hip holes: kangaroo (*Macropus* spp.) resting sites modify the physical and chemical environment of woodland soils. *Austral Ecology* **27**: 527–536.
50. Cogger, H. 2000. *Reptiles and Amphibians of Australia.* Reed New Holland: Sydney.
51. Dawson, T. 1995. *Kangaroos. Biology of the Largest Marsupials.* University of New South Wales Press: Sydney.
52. Baker-Gabb, D.J. 1988. The diet and foraging behaviour of the Plains-Wanderer *Pedionomus torquatus. Emu* **88**: 115–118.

5

What makes a good rocky outcrop?

In a nutshell

A rocky outcrop that has high conservation value for wildlife will typically have several or all of the following features:

- be large in size – typically 1 hectare or bigger – although smaller outcrops still provide habitat for rock-dwelling lizards and plants
- contain boulders and rock formations with crevices, vertical flakes, ledges, rock pools and caves
- a diverse overstorey of eucalypts, wattles and Kurrajongs as well as smaller trees such as Sandalwood and Quandong
- overstorey vegetation that is not too dense, as this can prevent sunlight from reaching the surface rocks, with negative effects on basking sites for reptiles
- a ground cover dominated by native plants including native tussock grasses
- appropriate fencing to exclude rabbits and livestock or control stock access and limit the risk of overgrazing and displacement of surface rocks
- be surrounded by remnant vegetation, tree plantings and native pasture rather than embedded in a heavily cleared cropping landscape.

Introduction

This chapter is about rocky outcrops, particularly small geological formations found in grazing and cropping landscapes. Rocky outcrops can vary in size from a few square metres to massive dome-shaped mountains covering many hundreds of

Box 5.1. The nature of rocks

Rocks can be classified into three groups according to their origin.

1. *Igneous rocks* form between 50 and 200 kilometres below the Earth's surface from molten rock (magma). Two types of magma are recognised: (a) basaltic magmas which flow over the landscape as lava; and (b) silicic magmas which are more viscous and cool below the Earth's surface to form granite-based rocks.
2. *Sedimentary rocks* form at the Earth's surface and originate from particles of eroded rock which are deposited on ocean floors or river beds. There are two broad types of sedimentary rocks: (a) rocks made of aggregates or fine mineral fragments; and (b) rocks derived from chemical precipitation of organic material such as limestone and chalk.
3. *Metamorphic rocks* are derived from igneous, sedimentary and other metamorphic rocks that are subjected to extreme temperature or pressure. There are many different kinds of metamorphic rock including quartzite, schist, gneiss and marble.

Over time, softer parts of the landscape erode leaving behind hard rocky outcrops. Water and wind continue to erode these rocks to form features that diverse and highly specialised plants and animals have evolved to depend on.

hectares. They can be formed by granite, basalt, schist and other rock types (see Box 5.1). Whatever type of rock they are made of, rocky outcrops share one feature in common; that is, they have an island-like appearance because they protrude from the surrounding landscape (Figure 5.1).

National parks and nature reserves in south-eastern Australia contain massive sedimentary and metamorphic rock formations. Examples include places renowned for plant species richness like the Grampians National Park (now called Gariwerd), as well as The Rock Nature Reserve, Cocoparra National Park and the conglomerate range of Benambra National Park (Figure 5.2).

Most of this chapter explores issues associated with small-sized (4–10 hectares) granite outcrops; however, areas of scattered surface rock and intrusions as small as a few isolated boulders are also important habitats on a farm. We have discussed this type of habitat in Chapter 2, 'What makes a good remnant?'.

Rocky outcrops are extremely important for wildlife on farms and they support a diverse range of species, particularly reptiles including skinks, geckos and large pythons.[1, 2] Granite outcrops are a common geological feature in south-eastern Australia but until recently their conservation values were poorly understood and they were rarely a priority for management. In this chapter, we discuss the common types of granite outcrops in rural landscapes and their ecological and conservation value. We conclude the chapter by identifying threatening processes and practical management actions that can be implemented to maintain outcrops.

Figure 5.1: A typical island-like granite outcrop in a farming landscape. (Photo by Damian Michael)

Figure 5.2: Tabletop Mountain, north of Albury in southern New South Wales, is part of a sedimentary conglomerate range. (Photo by Damian Michael)

Island mountains

Granite outcrops are typically called inselbergs in reference to their island-like appearance of rising abruptly from relatively flat landscapes.[3] Inselbergs can be classified according to their general size and shape. For example, steep-sided and dome-shaped formations are called bornhardts after a German geologist (Figure 5.3). Smaller inselbergs include formations such as castle koppies and nubbins characterised by either block-shaped boulder stacks or conical rock piles. Granite also forms scattered boulders called tors, ranging from loose clusters to entire hillsides covered in low embedded rocks. Granite shields, called pavements, protrude in flat landscapes, especially near Berrigan in southern New South Wales (see Figure 5.3). These main types of rock outcrop are common on farms and require management to maintain or enhance their habitat value for wildlife.

Ecological value of rocky outcrops

Flora

Rocky outcrops have a wide range of ecological values, many of which are not well recognised. The range of environmental and climatic conditions on rocky outcrops supports many different types of plants. In fact, studies in Western Australia have shown that granite inselbergs are exceptionally important ecosystems which support plants and vegetation communities found nowhere else in the world.[4] Granite outcrops on the South West Slopes of New South Wales support the only known populations of Granite Bushpea, Quandong, Sandalwood, Rock Correa and Woolly Ragwort in that region[5] (Figure 5.4).

Some of the species of flora that establish on the sun-exposed faces of rocky outcrops are lichens and resurrection plants. Resurrection plants shrivel during long periods without water but spring to life after rain, growing on sun-exposed rock faces. Mosses and succulents are commonly found in large mats on the sheltered parts of rocky outcrops. Plants such as these are collected by woodland birds like the White-throated Treecreeper to use as nesting material. Because steep-sided rocky outcrops are often inaccessible to livestock, they tend to support many native plants that are rare or absent in the surrounding landscape.

Fauna

Many animals are specialised outcrop-dwellers: crevices, slabs and vertical rock flakes provide habitat for species including the Snake-eyed Skink, Leaf-tailed Gecko (Figure 5.5) and the Flat Rock Spider. These animals have flat bodies, allowing them to shelter within narrow crevices. Deep cavities and rock shelves are occupied by large skinks, goannas and pythons (see Box 5.3).

Figure 5.3: (a) Bornhardt. (b) Castle koppie (c) Nubbin. (d) Tors. (e) Shield. (Photos by Damian Michael)

Rocky outcrops are crucial for the breeding cycle of butterflies. During the breeding season, male butterflies congregate in their hundreds atop rocky outcrops, waiting for a suitable mate. This behaviour is called hill-topping and occurs in a broad range of butterfly species.

Outcrops facilitate the movement of animals through landscapes both at a large and a small-scale. Large-scale movements include those migratory birds which use rocky outcrops as stopover locations. At a smaller scale of movement, rocky

Figure 5.4: Examples of rock-dwelling flora. (a) Woolly Ragwort. (b) Kurrajong. (Photos by Damian Michael)

Figure 5.5: The Leaf-tailed Gecko is a good example of specialised rock-dwelling species. (Photo by Damian Michael)

outcrops provide valuable stepping-stones for birds and bats. In the case of raptors, rocky outcrops help to create the thermal updrafts.

Rocky outcrops can have many ecological roles, making them some of the most species-rich parts of farms, particularly for reptiles

Rocky outcrops are valuable sources of water for animals. Rainwater that collects in rock pools (called gnammas) (Figure 5.6) not only provides animals with drinking water but supports a unique community of aquatic invertebrates.

Figure 5.6: Rock pool (gnamma). (Photo by Damian Michael)

Cultural values of granite outcrops

Indigenous people who found food, water and shelter in these places have a spiritual connection to outcrops. Evidence of this is found in the significant Aboriginal rock art on many outcrops. Outcrops also have post-European settlement cultural value, for example, because of the role they played in the era of the bushrangers. Morgan's Lookout near Walla Walla on the South West Slopes of New South Wales is a granite outcrop that was used as a refuge by bushranger Dan 'Mad Dog' Morgan (otherwise known as John Fuller (1830–1865)) (Figure 5.7).

Of course, granite outcrops are also valued for commercial (e.g. quarrying) and aesthetic reasons. The aesthetic beauty of granite outcrops is probably the main reason artists, bushwalkers and tourists are attracted to these places. Wave Rock in Western Australia and Mount Buffalo in north-east Victoria are prime examples of geological formations that are frequented by thousands of tourists each year. Other monolithic formations occur in most states of Australia (e.g. Bald Rock in northern New South Wales, Wilsons Promontory in Victoria, Porongurups in Western Australia, and Mount Wudinna in South Australia).

Attributes of rocky outcrops

Three key attributes of a rock outcrop influence their suitability as habitat for wildlife:

Figure 5.7: Morgan's Lookout – the hideout of famous bushranger, Dan 'Mad Dog' Morgan. (Photo by Damian Michael)

- the content of the outcrop, such as patch size and habitat complexity
- the condition of the outcrop, such as the level of disturbance and weediness
- the context of the outcrop, including topographical position and surrounding land use.

Below we discuss each of these three attributes in more detail.

Outcrop content

Outcrop content includes attributes such as size and habitat complexity. More reptile species occur as the size of an outcrop increases (Figure 5.8). A small outcrop covering a few square metres will usually support one or two species of reptile.[1] Outcrops of about 10 square metres in size can support approximately six species and large outcrops over 100 square metres can support more than a dozen species, including large pythons.[2] This relationship between outcrop size and reptile species diversity is due to the availability of additional kinds of habitats and extra space to accommodate more species. Large reptiles such as the Inland Carpet Python and Lace Monitor have home ranges that exceed 10–20 hectares. Outcrops less than 10 square metres are unlikely to support viable populations of these predators unless they are surrounded by large amounts of remnant vegetation.

Large rocky outcrops support more species of reptiles than small ones, but small outcrops are nevertheless important environmental and conservation assets on farms

Structural complexity can influence the number and type of animals found on an outcrop. This relates to the amount of different types of habitat that are formed as boulder stacks increase in height.

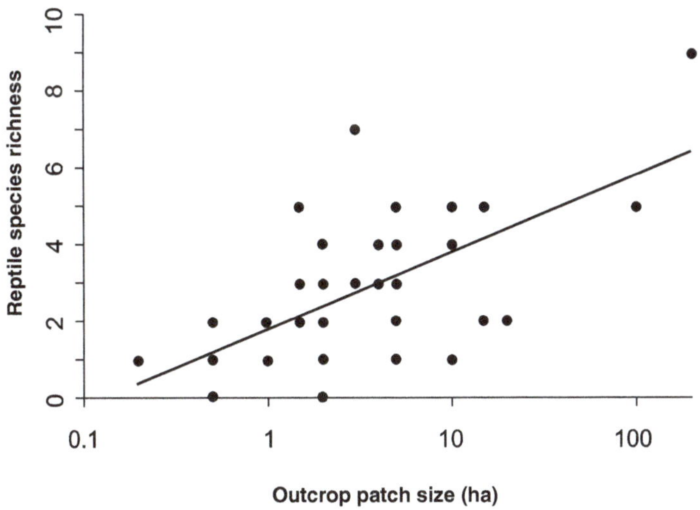

Figure 5.8: The relationship between reptile species richness and granite outcrop patch size in the South West Slopes of New South Wales.

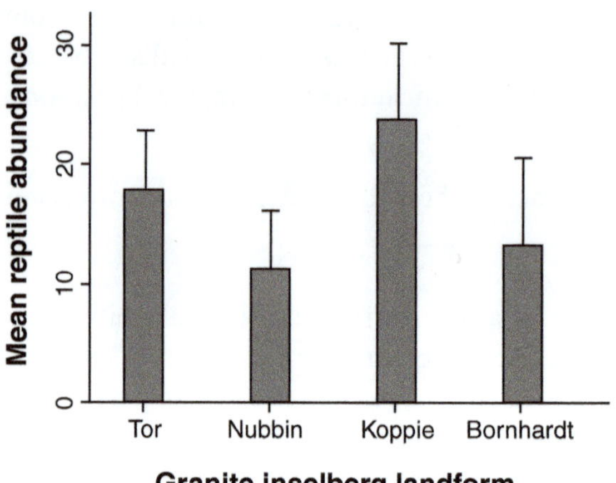

Figure 5.9: Changes in total reptile abundance among four structurally different types of granite inselbergs in the South West Slopes of New South Wales.

Inselbergs vary in the amount and arrangement of boulder formations and surface rock. The most structurally complex landforms such as castle koppies are the kinds of outcrops that can provide habitat for large populations of rock-dwelling species of reptiles (Figure 5.9). Outcrops with large boulders and many crevices support more and larger families of the Crevice Skink (see Box 5.2).

Box 5.2. The Crevice Skink – a social rock-dweller

The Crevice Skink is a medium-sized lizard that produces from two to six live young during January and February. It feeds on insects and plant material and shelters within rock crevices, behind bark, and within the cracks of tree stumps and fallen timber. It is one of the most common species of lizards on granite outcrops on farms, even on granite outcrops completely devoid of native vegetation. One reason they are so common is their partly herbivorous diet and the fact that they excrete small scat piles on rock ledges which attract a range of insects which the lizards then eat.

The Crevice Skink forms small family groups on rocky outcrops that usually consist of a breeding pair and several age cohorts of their offspring. On any given rock outcrop, there may be several different family groups. Family groups spend most of their time basking on rock faces or sheltering within narrow crevices. When disturbed, they will retreat within a crevice but will often reappear to investigate the intruder. They are territorial and will aggressively defend their homes from other lizards, including non-family members of their own kind. Large extended family groups and a greater number of families occupy the most structurally complex outcrops and the highest boulders with lots of crevices. Lizards found near the edge of the outcrop are often sub-adults and are solitary.

Figure 5.10: Crevice skink in crevice. (Photo by Damian Michael)

Outcrop condition

Disturbances which affect the habitat value of a rocky outcrop include: (1) logging; (2) overgrazing by domestic livestock, rabbits and hares; (3) weed invasion; and (4) excessive quantities of soil nutrients. The type of disturbance regime varies depending on where an outcrop is located in the landscape. Outcrops in grazing landscapes are less degraded than outcrops in cropping landscapes. Understanding how wildlife responds to different kinds of disturbance is an important part of developing appropriate management strategies for outcrops. For example, reptile populations can be influenced by the amount of exotic grass cover on an outcrop (Figure 5.11). This is because critical habitat such as bare ground and basking sites are eliminated when exotic grasses colonise an outcrop.

Dense regrowth vegetation also can reduce reptile numbers on rocky outcrops (Figure 5.11). This is because reptile basking sites are shaded by a dense tree canopy. Grazing pressure also can reduce native ground cover and increase unpalatable weeds which contribute to the shading of surface rocks. Sheep dung also can accumulate beneath rocks and make them unsuitable shelter sites for reptiles and invertebrates. High-intensity grazing, particularly set-stock grazing, can have a significant negative effect on reptiles.[5]

Vegetation condition is one of the key factors influencing rocky outcrops. Careful management can significantly improve outcrop condition for reptiles as well as other groups, including native plants

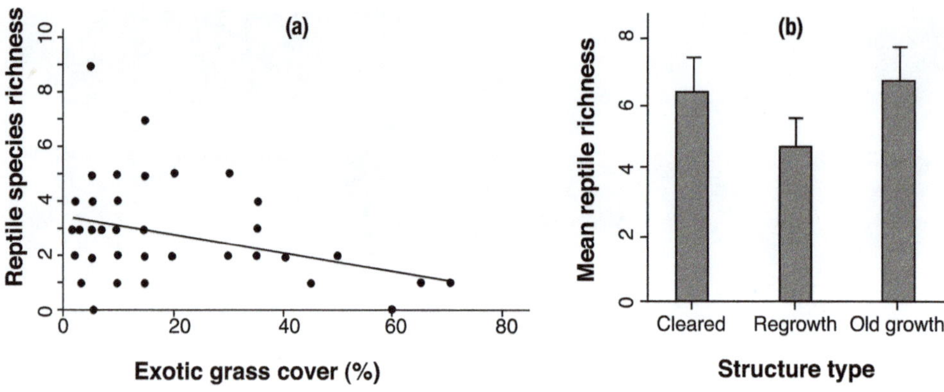

Figure 5.11: The relationship between reptile species richness and: (a) exotic annual grass cover, and (b) overstorey vegetation structure in the South West Slopes of New South Wales.

Outcrop context

The plants and animals occurring on an outcrop can be significantly influenced by the kind of land use in the surrounding areas. For example, as surrounding land becomes increasing modified, there is a reduction in reptile diversity. Rocky outcrops in intensive cropping landscapes are often exposed to chemical spray drift and high rates of weed colonisation. Intensively cleared landscapes also create a barrier to organisms attempting to disperse between outcrops. This can lead to inbreeding and local extinctions if populations cannot be replenished by immigrants arriving from other outcrops or from adjacent habitats. Some reptiles, such as Cunningham's Skink, avoid inbreeding by forming monogamous pairs and stable family groups. Recognition of closely related family members is one way this long-lived species avoids mating with family members.

Threats to rocky outcrops and their management

Threats to granite outcrops come mainly from agriculture. Feral animals and commercial bush-rock removal also can negatively affect the biodiversity value of rocky outcrops.[6] In many cases, granite outcrop management is relatively simple and can include fencing to exclude livestock, halting timber removal, and allowing native vegetation to recover. In some cases, outcrop management may require other approaches like thinning to reduce densely spaced regrowth stems. We discuss these threats and management actions below.

The major threats to rocky outcrops can be controlled by managing fire, fencing, excluding livestock or controlling grazing pressure, halting timber removal, and allowing native vegetation to recover

Altered fire regimes and fire management

Fires are generally rare events on rocky outcrops; however, frequent, high severity fires can alter vegetation structure and change plant species composition (see Figure 5.12). Many plants growing in rocky areas are unable to resprout after fire. These species are called obligate seeders and can be eliminated by frequent, high intensity wildfires. Changes to vegetation structure and plant species composition also can have negative impacts on wildlife such as by altering the amount of sunlight reaching the basking sites of reptiles.

Lightning strikes are the most common cause of fire on rocky outcrops, but human-induced fires are becoming increasingly common in agricultural areas. Responses to controlling fire on rock outcrops often involve lighting a perimeter back-burn. The use of very high severity back-burns can have major negative environmental effects and their use is to be avoided wherever possible. Perimeter back-burns can have at least five kinds of detrimental effects on rocky outcrops and their associated plants and animals:

- Perimeter fires are usually ignited at the base of the outcrop causing the fire to travel uphill and burn with greater severity than a fire travelling downhill.
- Native animals can become trapped and killed by high severity, fast-moving perimeter back-burns.
- The amount of heat generated during back-burning can completely consume hollow logs, and large standing dead trees.
- High severity perimeter back-burns can kill seed stored in the soil. Subsequent widespread removal of native vegetation allows agricultural weeds to colonise bare ground. Loss of native vegetation also increases soil erosion and nutrient runoff.
- Fragile rock formations (vertical flakes and exfoliations) can be destroyed by high-severity perimeter back-burns and it may take several hundred years for them to be replaced.

Grazing regimes and livestock management

In recent years, there has been a focus on developing grazing management strategies to enhance biodiversity conservation outcomes in production landscapes.[7] Too much grazing can degrade outcrop environments. Too little grazing can lead to regrowth vegetation significantly altering light penetration to the ground with corresponding effects on reptile basking sites.

Two kinds of grazing strategies may be useful in managing the fringing vegetation on the flared slopes of an outcrop. These are: (1) crash grazing, involving high stocking levels for a short period of time to reduce biomass and minimise the

Figure 5.12: (a) Obligate seeders (Senecio, Isotome) after (b) a back-burn on Stringybark Hill near Albury in southern New South Wales. (Photos by Damian Michael)

Figure 5.13: Sheep camping on a rocky outcrop. (Photo by Damian Michael)

selective consumption of desirable plant species; and (2) time-controlled grazing, involving grazing during specific stages of a plant's development, such as during flowering. Both strategies require long-term planning and careful monitoring. If appropriate grazing regimes are applied, then a balance between production and conservation may be achieved. Indeed, grazing could be a useful tool in controlling dense regrowth vegetation and broad-leaved exotic plants.

Woody weeds and other invasive exotic plants

Approximately 45% of the plants associated with granite outcrops on the South West Slopes of New South Wales are introduced species.[5] This can have negative effects on rock-dwelling animals (Figure 5.11). Rock outcrops such as scattered tors are often dominated by introduced agricultural grasses like Barley Grass, Rye Grass and Wild Oats, whereas bornhardts support relatively few agricultural species. Instead, bornhardts often contain a higher proportion of plants unpalatable to livestock like Inkweed, Blackberry Nightshade, Sweet Briar and Prickly Pear.

Granite outcrops that are close to towns often contain a high proportion of plants that escape from nearby gardens or are spread by the introduced European Blackbird. Some examples of woody weeds that commonly invade these kinds of rocky outcrops include Privet, Cotoneaster, Boxthorn, Firethorn and Cherry Plum.

In most cases, woody weeds can be cut and poisoned using a glyphosate-based herbicide. Local councils also should be encouraged to develop strategies to minimise plant invasions. This might mean placing restrictions on the types of plants that can be grown in areas adjacent to bushland, a challenge urban planners should consider when new developments are proposed. Broad-leaved weeds compete with native forbs for space and nutrients and have the potential to reduce native invertebrates and foraging areas for reptiles and woodland birds. Both ecological fire regimes and the grazing methods which we mentioned above may be useful management tools in controlling these kinds of invasive species. Reducing nutrient levels by restricting the use of super-phosphate also can help reduce broad-leaved weeds.

Pest animal control

The Red Fox and the Feral Cat are significant predators of Australian wildlife. They prey on a wide variety of native animals including native mice, small marsupials, and reptiles. Predation of native animals inhabiting rocky outcrops can be substantial and may eventually lead to localised extinctions. As an example, the Inland Carpet Python is a species that breeds every three years in southern New South Wales and occurs in low population densities. It has declined considerably in recent years due to habitat loss as well as predation by the Red Fox.[8] Fox control programs with the most significant positive effects on wildlife populations are those conducted regularly across several adjoining properties (see Chapter 7).

The European Rabbit can have a significant effect on the distribution and abundance of rock-dwelling flora and fauna by reducing vegetation cover and promoting soil erosion. This introduced species can reach high population densities on rock outcrops where it can be extremely difficult to eradicate. One method traditionally used to reduce rabbit populations was to block the entrances of rock crevices and cavities with wire-mesh. But this method proved ineffective as rabbits simply found alternative places to dig their burrows. Today, quite a few granite outcrops still contain the remnants of wire-mesh. It should be removed to prevent injury to native animals, particularly those species that regularly use rock crevices like the Carpet Python. Rabbit control in areas with granite outcrops requires strategic population control methods (see Box 5.3).

The introduced Portuguese Millipede is a common pest invertebrate found beneath surface rocks on granite outcrops. A study in the Warby Ranges National Park near Wangaratta found that millipedes had a significant negative effect on rock-inhabiting mosses and lichens.[9] A reduction in these plants can lead to increased water runoff and excessive soil erosion. The Portuguese Millipede is thought to be increasing in abundance and distribution, but the full impact of this species is still unknown. Further research and monitoring is needed to determine the impacts of this species on rock-dwelling fauna, and to guide effective control methods.

Box 5.3. Controlling rabbits and conserving pythons – a true management challenge

The Inland Carpet Python or Murray Darling Carpet Python is a non-venomous snake that grows to over three metres in length. It preys on mammals and birds and is found in a broad range of habitats throughout southern Queensland, inland New South Wales and northern Victoria. Rabbits are their main food source, but pythons will also venture into roof cavities, grain silos and hay sheds in search of rats and mice.

The Inland Carpet Python was once relatively common, especially during the rabbit plague years prior to the 1950s and before CSIRO released the viral disease myxomatosis to control rabbit populations. When rabbits began to decline, so too did the python. Today, the Carpet Python is a species of conservation concern and in southern New South Wales it is only found along vegetated river systems and on large rocky outcrops.[2] Granite outcrops in good condition provide excellent habitat for pythons, but even partly cleared granite outcrops can support python populations. During the winter months, pythons become inactive and spend around four months sheltering within tree hollows or rock crevices. During the warmer months, pythons spend most of their time sheltering within rabbit burrows. Therefore, in the absence of small and medium-sized burrowing marsupials (which are extinct in southern New South Wales), the future conservation of python populations will depend on strategic rabbit control. For example, warren systems should be ripped and fumigated only during cooler months when pythons are less likely to be using them as shelter. Burrow implosion techniques should not be used on granite outcrops to avoid permanent damage to shelter-sites.

Figure 5.14: Carpet Python. (Photo by Damian Michael)

Figure 5.15: Portuguese Millipede. (Photo © CSIRO)

Granite outcrop restoration

Damage to rock formations and fragile microhabitats such as flakes and exfoliations can have a long-lasting effect on rock-dwelling plants and animals. Permanent damage can be caused by illegal reptile collectors, rock-climbers, trampling by stock, extreme fire events, quarrying activities and bush-rock collectors. Bush-rock removal is a threatening ecological process but can be extremely difficult to regulate and in many cases mossy rocks are extracted from privately owned bushland. Ideally rocks should be sourced from existing quarries. Anyone wishing to use rocks to decorate gardens should ask their local nursery or rock supplier where the rocks come from.

Recreational activities – rock-crawling

'Rock-crawling or rock-hopping' is an extreme off-road American motor sport which is gaining popularity in Australia. Purpose-designed and constructed dune-buggies are craned on top of rock outcrops where drivers then have to negotiate a course which involves scaling large boulders and rock piles. Rock-crawling can cause substantial damage to rock formations and vegetation. Once these microhabitats are lost they are not easily replaced.

Revegetation programs

Revegetation programs are becoming increasingly common in agricultural landscapes (see Chapter 3). Plantings near rocky outcrops should be designed carefully so they do not compromise the quality of these environments for rock-dwelling flora and fauna. There is potential for tree plantings to have a negative effect on some rock-dwelling species by shading basking sites and foraging areas. Tree plantings on or around rocky outcrops should incorporate the following design principles.

- *Plant density and spacing patterns.* Trees should be planted at a density of 20–30 per hectare to reflect the natural spacing patterns in grassy woodlands.
- *Plant species composition.* Granite outcrops often support a diverse range of shrub species of different sizes. Therefore, plantings should incorporate species that vary in height and composition. Plantings also should support species that occur naturally on rocky outcrops including Quandong, Sandalwood, Kurrajong and Native Pine (*Callitris* spp.). Plantings also should contain native grasses. Kangaroo grass seed mulch can be an effective way of restoring swards of native grass. The ground may need to be prepared (e.g. spraying herbicide) before the mulch is applied.
- *Planting aspect.* Reptiles are usually more common on the sunniest parts of rock outcrops. If an outcrop is dome-shaped or conical, then this will be the eastern, northern and western aspects. Plantings established on the southern side of a rock outcrop will be less likely to shade important basking sites.

Figure 5.16: Three-toed Skink. (Photo by Damian Michael)

- *Incorporate additional structure.* Fallen timber that might otherwise be burnt can be relocated into tree plantings on outcrops. Similarly, old fence posts and other wood off-cuts can be placed in plantings to provide shelter sites for small lizards and invertebrates. Dead shrubs should be retained as this type of habitat is commonly used by the Olive Legless Lizard.
- *Control livestock grazing.* Tree plantings where livestock is excluded have higher reptile diversity than grazed plantings. Some reptiles such as the Olive Legless Lizard and the Three-toed Skink are more common in ungrazed tree plantings than in grazed remnant vegetation. This is because perennial grass that is left to mature is an important habitat that is usually limited in agricultural landscapes.

Summary

Rocky outcrops are critically important environments for native plants and animals in agricultural areas. They support many specialist species that are rare or absent elsewhere in farming landscapes. Rocky outcrops need special management attention to maintain their ecological integrity. Key management practices include management of grazing pressure by domestic livestock, appropriate fire management, carefully implemented feral animal control programs, fencing, appropriate planting programs, and weed control.

Box 5.4. Beelawong – a case study for managing rocky outcrops for biodiversity conservation

Near the town of Gerogery, in southern New South Wales, Chris and Sue Cain of 'Beelawong' are managing a large granite outcrop to improve the condition of native vegetation and the conservation of wildlife. In 1997, they constructed a fence around the base of the outcrop to exclude livestock and commenced an invasive animal and plant control program.

Poison baits are laid on the property and on the outcrop on a regular basis to reduce fox numbers and minimise predation of the Inland Carpet Python, a species of conservation concern which occurs in low population densities on rocky outcrops in the region. The Cains work closely with their neighbours and the Livestock Health and Pest Authority (formerly the Rural Lands Protection Board) to control rabbits on their outcrop. This involves laying poison baits instead of ripping or imploding rabbit burrows. Invasive exotic plants such as St Johns Wort are spot-sprayed to help native ground cover plants to recover.

The management of the Cain's outcrop is now relatively passive. Since grazing was removed, native vegetation has begun to naturally regenerate and today the rocky remnant supports a diversity of rare and threatened plants.

Figure 5.17: Well-managed rocky outcrop at Beelawong. (Photo by Damian Michael)

References

1. Michael, D.R., Cunningham, R. and Lindenmayer, D.B. 2008. A forgotten habitat? Granite inselbergs conserve reptile diversity in fragmented agricultural landscapes. *Journal of Applied Ecology* **45**: 1742–1752.
2. Michael, D.R. and Lindenmayer, D.B. 2008. Records of the Inland Carpet Python, *Morelia spilota metcalfei* (Sepentes: Pythonidae), from the south-western slopes of New South Wales. *Proceedings of the Linnean Society of New South Wales* **129**: 253–261.
3. Twidale, C.R. 1982. *Granite Landforms*. Elsevier: Amsterdam.
4. Porembski, S. and Barthlott, W., eds. 2000. *Inselbergs: Biotic Diversity of Isolated Rock Outcrops in Tropical and Temperate Regions*. Springer-Verlag: Berlin.
5. Michael, D.R. 2008. Insular granite outcrops: botanical refuges in agricultural landscapes. *Woodlands Wanderings: Newsletter of the Grassy Box Woodland Conservation Management Network, Canberra* **6**: 4–6.
6. Michael, D.R., Lindenmayer, D.B. and Cunningham, R.B. 2010. Managing rock outcrops to improve biodiversity conservation in Australian agricultural landscapes. *Ecological Management and Restoration* **11**: 43–50.

7. Dorrough, J., Yen, A., Turner, V., Clarke, S.G., Crosthwaithe, J. and Hirth, J.R. 2004. Livestock grazing management and biodiversity conservation in Australian temperate grassy landscapes. *Australian Journal of Agricultural Research* **55**: 279–295.
8. Heard, G.W., Robertson, P., Black, D., Barrow, G., Johnson, P., Hurley, V. and Allen, G. 2006. Canid predation: a potentially significant threat to relic populations of the Inland Carpet Python *Morelia spilota metcalfei* (Pythonidae) in Victoria. *The Victorian Naturalist* **123**: 68–74.
9. Smith, M. 2003. Environmental damage by *Ommatoiulus moreleti* (the Portuguese Millipede) in the Warby Range, Victoria. Hons thesis. La Trobe University: Wodonga.

6

What makes a good waterway?

In a nutshell

A good waterway will typically have several or all of the following features:

- appropriate fencing to either exclude or control stock access and limit the amount of grazing and trampling pressure
- areas of native vegetation including large trees, understorey trees and shrubs, and native ground cover
- areas of native vegetation that extend into a water body, including wetland vegetation and emergent water plants
- places that vary in the rate of stream flow and vary in water depth
- some submerged or semi-submerged fallen logs, known as snags, and branches
- protection from fire and chemical spray drift.

Introduction

The wet areas on a farm can include streams and creeks, natural wetlands, farm dams and irrigation channels. Significant amounts of biodiversity can be associated with these areas. Like all of the other key features of farms, how waterways are managed can have a highly significant effect on their suitability for wildlife.[1] This chapter focuses on the attributes of a good waterway on a farm. Our discussion encompasses streams and creeks, natural wetlands, and farm dams. Most of the discussion is on terrestrial animals (particularly birds) that use streams, wetlands, farm dams, and irrigation channels. We are acutely aware of the critical importance of these environments for aquatic animals like fish and invertebrates, but we do not discuss them in detail in this chapter.

Why are waterways so important for farm biodiversity?

Water is essential for all life in temperate woodlands. Access to good quality water is also essential for livestock grazing. Many studies have demonstrated the ways in which waterways are vital for wildlife. A short list of some of these is set out below:

- Bird diversity is often higher around streams and gullies than elsewhere in the landscape[2, 3] including in plantings established in these areas (see Chapter 3).
- The bird assemblages found around streamlines and riparian vegetation are different from those in the rest of the landscape.[3, 4]
- Some species of terrestrial birds are more likely to occur in remnants located close to waterways and farm dams.
- Exposed earth banks around waterways provide important breeding areas for birds such as the Rainbow Bee-eater, Welcome Swallow, Fairy Martin and Spotted Pardalote.
- Population densities of native mammals like the Squirrel Glider and the Bush Rat are greatest in riparian areas.
- Mammals such as the Common Ringtail Possum have more young in gullies.[5]
- Riparian areas appear to act as dispersal routes for native mammals such as the Bush Rat.[6]
- Animals like Myotis bats forage directly above and at the water surface.
- The size and density of trees and shrubs and the number of layers of vegetation is often greatest around waterways – because of the additional moisture and nutrients that characterise these areas. Waterways also support a specialised flora, with many plant species being largely confined to them.

The wet parts of a farm – rivers, creeks, wetlands and farm dams often support more plants and animals and different kinds of plants and animals than other parts of a farm

Attributes of well-managed streams and natural drainage lines

Well-managed streams and natural drainage lines can be particularly valuable for wildlife if they have two key characteristics. These are briefly outlined below.

Well-developed areas of riparian vegetation

A good waterway has vegetation on its banks (i.e. not extensive areas of bare banks). This includes vegetation that extends from drier areas into the water,

Figure 6.1: Rainbow Bee-eater – a species that nests in earth banks, especially in riparian areas. (Photo by Suzi Bond)

including wetland vegetation and emergent water plants.[7] Riparian vegetation can comprise large trees, understorey trees and shrubs and native ground cover. This vegetation provides places for animals to forage and to breed. For example, many species of frogs need riparian trees and shrubs in which to successfully reproduce. Riparian vegetation also plays many other key ecological roles. For example, it: (1) filters nutrients and sediment from the surrounding terrestrial environment; (2) provides shade for in-stream areas to limit water temperatures that might otherwise be fatal for fish and other aquatic organisms; (3) acts as a source of nutrients to aquatic ecosystems through leaf fall and bark shedding; and (4) provides a source of logs and branches that fall into waterways. Submerged and semi-submerged logs and branches are nesting and foraging places for many animals ranging from fish to invertebrates.

Native vegetation around, at the margins and within creeks, wetlands and farm dams provide valuable habitat for many species

Areas of slow-moving water

A good waterway is not a fast-moving stream along its entire length. Rather, it should be characterised by deep pools, chains of ponds and shallow areas of fast-flowing water such as ripple zones. This combination of features is crucial for

Figure 6.2: (a) A new generation of River Oaks and River Red Gums along the Murrumbidgee River. (b) A well developed area of riparian plantings. (Photos by Rebecca Montague-Drake)

Figure 6.3: An artificial groyne build into the Tharwa River (ACT) to slow stream flow and create deep pools needed by the endangered Trout Cod. (Photo by Mark Lintermans)

the persistence of some species of frogs. It is also essential for an array of species of native freshwater fish, as well as mammals like the Platypus.

Threats and their management

A range of factors threatens the integrity of streams and natural drainage lines on farms. Two key ones are the extensive removal of riparian vegetation, and overgrazing and vegetation trampling by domestic livestock. These can result in highly elevated levels of nutrients in waterways with corresponding negative impacts on water quality.

Removal of native vegetation and overgrazing can have other negative impacts, such as increased erosion with subsequent sedimentation and silting of waterways in some parts of streams, whereas in others it can result in streambank incision, leading to faster rates of water flow and loss of chain-of-ponds systems (see Figure 6.5). Linking formerly disconnected chains of ponds can have devastating effects on frogs because: (1) the increased speed of water flow can sweep away their eggs; and (2) they become more susceptible to introduced predators like the Mosquito Fish.[8]

Halting vegetation removal and controlling grazing pressure are the two most important ways to protect the quality of habitats associated with creeks, streams and farm dams

Figure 6.4: While individual trees have been protected from stock by tree guards, uncontrolled access to water by stock is causing severe erosion at this site. (Photo by Rebecca Montague-Drake)

Some simple management actions have a significant positive effect on waterway condition, the quality of water for livestock, and the quality of streams and natural drainage lines as habitat for farm wildlife. First, it is important to halt or greatly limit the amount of clearing of riparian vegetation. Second, fencing and the controlled access of livestock to waterways can allow native vegetation to recover, including submerged and semi-submerged aquatic plants. Third, riparian areas can be good places to target for active planting programs to restore native vegetation cover (see Chapter 3).

Fire poses a threat to the vegetation in natural drainage lines such as wetlands. These areas can dry out during prolonged droughts and then be susceptible to inappropriate burning practices. The vegetation and the fauna typical of natural wetland environments can be extremely sensitive to fire and may take a prolonged period to recover, especially if the burned area is subsequently grazed by livestock. This is, in part, because the vegetation and the fauna has to deal with the cumulative effects of three kinds of disturbance in rapid succession – prolonged drought, fire and grazing.

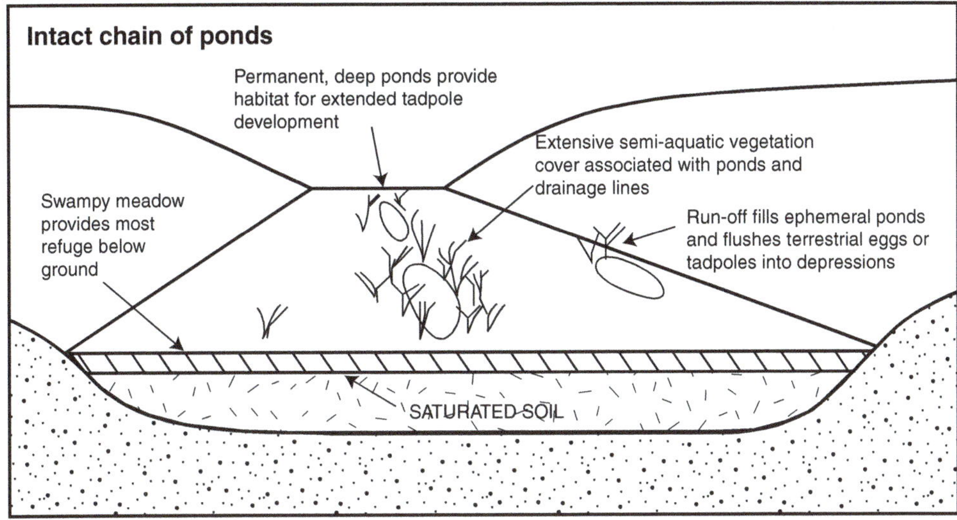

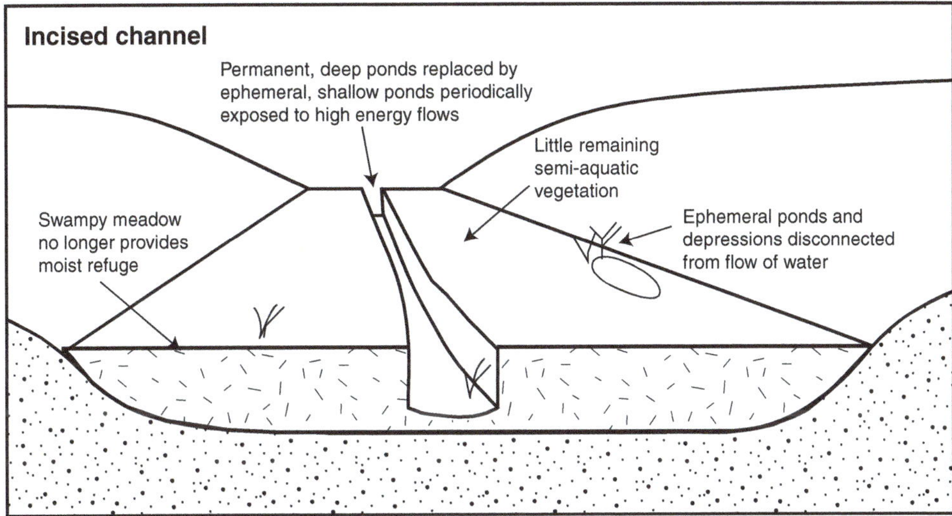

Figure 6.5: Changes in a stream environment comprised a chain-of-ponds to an incised stream bed resulting from vegetation clearing and poorly managed streams.

Attributes of well-managed farm dams

Farm dams are bodies of still water. Water will often flow into them, but not from them, except during floods. Many people do not associate farm dams with native wildlife, but they can be important environments for frogs and birds. For example, bird species richness is higher on farms with dams than those where they are absent. Even the flat muddy areas around farm dams are important environments

Box 6.1. Restoration programs in riparian areas – use natives not exotics

Many restoration programs target gullies and riparian areas. These revegetation works can help tackle problems with soil erosion and salinity. They also can be important for native wildlife – more species of birds are likely to occur in riparian areas that have been replanted than plantings established elsewhere on a farm.[9] Some of the unusual birds that can occur in plantings in riparian areas include the Dollarbird and the Yellow-tufted Honeyeater. Creating suitable habitat for these native birds and many other native animals requires that native plants are used in revegetation programs. Conversely, exotic plant species are another factor which can degrade the integrity of waterways on farms. For example, willows can change streambank conditions in ways that make them unsuitable for bank-nesting species like the Platypus. Where willows infest creek banks they crowd out native reeds, rushes and shrubs and become unsuitable habitat for frogs, native insects and small fish. Exotics trees such as the willow should never be established on any part of a farm, including around waterways. Indeed, active and well-targeted willow removal programs have commenced in many areas. To maintain stable bank conditions, it is important to replace these exotic trees with native vegetation cover, as well as retain the stump and root ball of cut trees.

Figure 6.6: Planted native vegetation in a riparian area. (Photo by David Lindenmayer)

Figure 6.7: (a) Willows crowd out native plants and change streambank conditions, making them unsuitable habitat for animals such as frogs, fish and Platypus. (b) Willow removal programs are being instigated in many areas but it is important to restabilise the streambank by planting native species following willow removal. (Photos by Rebecca Montague-Drake)

Box 6.2. The biodiversity benefits of removing willows

On the Murrumbidgee River at Gundagai, New South Wales, the local anglers' club, Bushcare group and the council have removed willows from large stretches of the riverbank with assistance from the Murrumbidgee Catchment Management Authority. Within six months of poisoning willows and excluding stock, native reeds and rushes recolonised the area. The return of these habitats resulted in dramatic increases in frogs, birds, insects and not only small fish but also large fish such as Yellow Belly which take advantage of the bountiful supply of food now available in this part of the river.

– often being used as foraging sites by such species as the Black-fronted Dotterel. The mud from these areas is also used by the White-winged Chough, Magpie Lark and Welcome Swallow for nest building.

As in the case of streams and natural drainage lines, there are some important factors which strongly influence both the quality of the water in farms dams and also their value as habitat for farm wildlife (see Figure 6.11). A key one is native vegetation – around the edges of a farm dam, at the boundaries of the water line, and in the water. Native vegetation that surrounds a farm dam shades the water, reduces rates of evaporation, and lowers water temperatures, making algal blooms less likely to occur. It also filters sediments and excess nutrients (e.g. from animal dung), thereby contributing positively to water quality – both for livestock and native animals and plants. The native vegetation surrounding a farm dam is used

Box 6.3. Re-snagging waterways

The loss of large logs from streams has long been recognised as a key process threatening the integrity of waterways in many parts of Australia. These losses can occur in two ways – by deliberate removal, known as de-snagging, and by the clearing of riparian vegetation which removes the source of logs and large branches. Significant steps are being taken in some regions to address the problem of the lack of logs in streams. For example, the New South Wales Roads and Traffic Authority, in partnership with the Murray–Darling Basin Authority, has embarked on a major project to re-snag many kilometres of the Murray River. The source of logs is the large number of trees cut down as part of the expansion of the Hume Highway in southern New South Wales. The aim of this substantial river restoration program is to recreate native fish habitat and improve the overall ecology of the river. Snags are critical habitat for native fish such as Murray Cod which use them for breeding, sheltering and as foraging sites. Snags not only provide important habitats for a diverse range of aquatic life, they also play an important role in shaping rivers and creeks, creating variation in the depth and flow of a waterway, resulting in a diversity of habitats for wildlife.

Figure 6.8: Snagged area of river in southern New South Wales. (Photo by Rebecca Montague-Drake)

Figure 6.9: Burning area of natural wetland near Gundagai. (Photo by David Lindenmayer)

Box 6.4. Environmental wetting of Black Box and Red Gum

Australian rivers and creeks are characterised by enormous variation in levels of stream flow. In fact, stream flows are more variable in this country than anywhere else on the planet. Large numbers of native species are well adapted to flooding and drying cycles. Therefore, major environmental problems can arise when flooding and drying regimes are altered. Native vegetation can die and its associated biodiversity is lost when areas are either permanently inundated or deprived of intermittent flood water. This has occurred in extensive areas of Black Box and River Red Gum woodland in southern New South Wales and northern Victoria. As part of efforts to address this problem, management programs have commenced to reinstate natural wetting and drying cycles in areas like the lower Murray Catchment in southern New South Wales. These management efforts are a key part of maintaining woodland tree health, stimulating tree regeneration and enhancing the condition of areas of natural wetlands.

for breeding by native animals (e.g. frogs), as well as for perching, calling and foraging by birds. For example, the cover of native reeds like Cumbungi around the margins of farm dams provides habitat for the Australian Reed-warbler (see Box 6.5), as well as protecting it from potential predators.

Figure 6.10: The native Spotted Marsh Frog commonly uses farm dams for breeding. (Photo by Damian Michael)

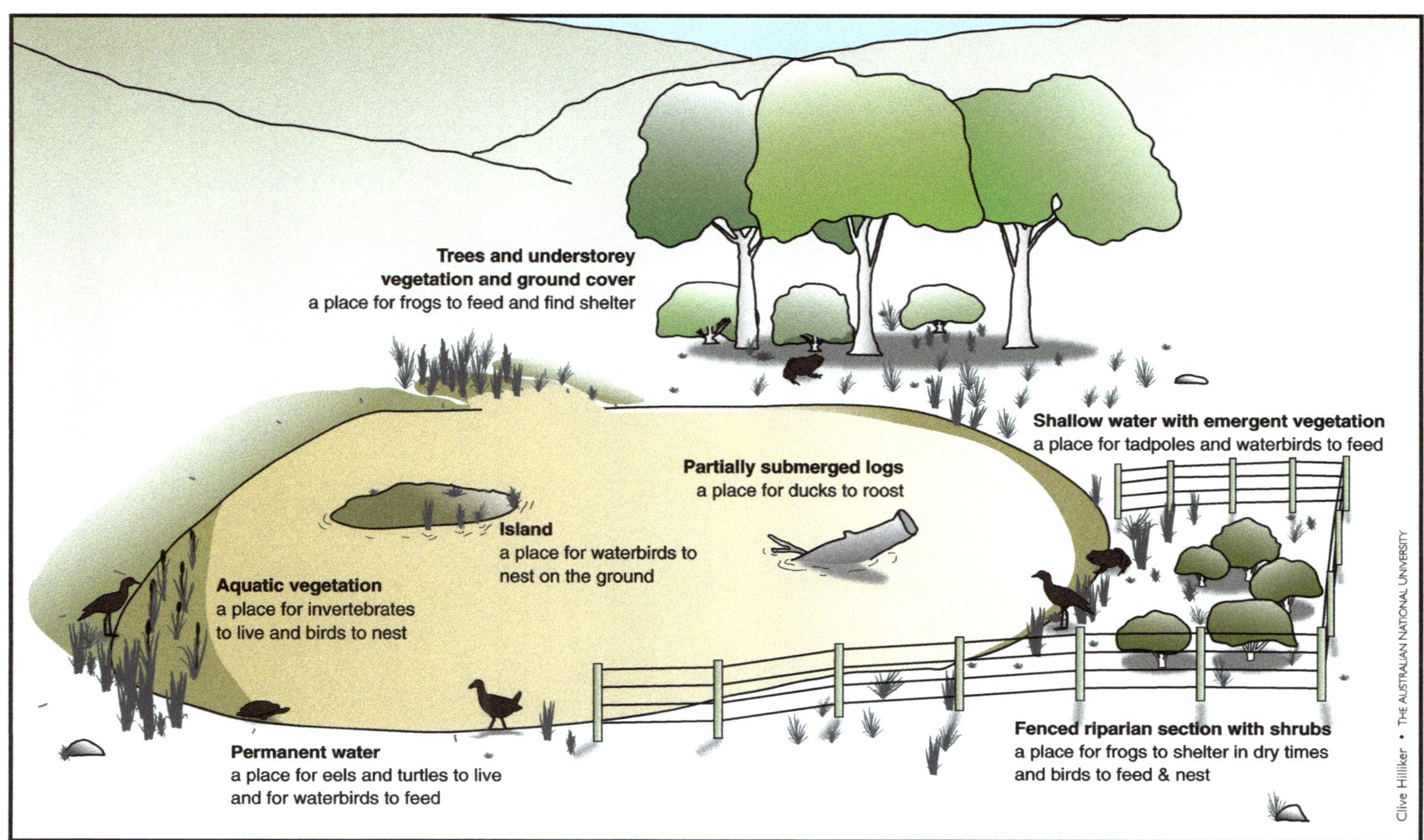

Figure 6.11: Diagram showing the key features of a well-managed farm dam. (Drawing by Clive Hilliker)

Figure 6.12: The contrast between: (a) a well-managed farm dam, and (b) a poorly managed farm dam. (Photos by David Lindenmayer)

Figure 6.13: The Australian Reed-warbler requires farm dams with well-developed reed beds at the water margin. This bird species has a classic and distinctive call of 'warty-warty-warty-creep-creep-creep'. (Photo by Julian Robinson)

The presence of submerged and semi-submerged vegetation in a farm dam can make it a suitable nesting environment for waterbirds like the Hoary-headed Grebe and Australasian Grebe. This kind of vegetation also influences the suitability of dams for a range of species of frogs.

Box 6.5. Bullrushes /Cumbungi (*Typha* spp.) – wonderful or weed?

In books on invasive plants in Australia, reeds and rushes, but more particularly Cumbungi, are often identified as weeds. The majority of these plants are native and are important in the ecology of creeks, rivers and wetlands. Given this, are Cumbungi and other reeds and rushes a weed or an asset? The reason why these plants are mentioned in many books on weeds is that they grow in irrigation channels. In such situations, these plants (particularly Cumbungi) can interfere with the hydrological performance of these structures; however, in natural waterways and farm dams, Cumbungi and similar plants play essential ecological roles. These include: (1) maintaining water quality – filtering excessive salts, nutrients, sediments and other pollutants; (2) slowing erosion – reeds stabilise the soil on the water edges and creek bed, and also slow flows, reducing erosion; (3) restoring degraded waterways – the re-establishment of reeds and rushes in creeks and rivers can assist in reshaping our watercourses to a more natural shape and hydrology, recreating what is often referred to as a chain-of-ponds system; and (4) providing habitat – reeds and rushes are essential habitat for many species of terrestrial and aquatic wildlife. If an aim is to reduce the dominance of reeds and rushes, particularly Cumbungi, it is possible to inhibit their growth by planting native trees and shrubs.

Figure 6.14: Bullrushes/Cumbungi. Three species of Cumbungi occur in Australia: Narrow Leaf Cumbungi *Typha domingensis* and Broadleaf Cumbungi *Typha orientalis* (pictured), both of which are native and *Typha latifolia* which is an exotic species that is now found in some parts of Victoria. (Photo by John Tann)

Threats and their management

The removal of native vegetation around farms dams can have a significant impact on both water quality and wildlife habitat suitability.[1] This problem can be compounded by unregulated livestock grazing and/or uncontrolled stock access to water. The way to address this problem is to halt or limit native vegetation clearing around farm dams and replant appropriate native vegetation where cover is limited. In addition, gates and fences can be used to control stock access to water or water can be piped to nearby troughs.

Two other potential threats to wildlife around farm dams can be readily managed. First, the use of barbed wire in fences around farm dams can snag and kill flying or gliding animals. For those fences where it is not feasible to exchange barbed wire for straight wire, it can be useful to enclose barbed wire in a poly-pipe covering or increase its visibility by tying flagging tape to the barbed rung. Second, the removal of weeds is often needed as part of restoring waterways on farms. Many herbicides (e.g. products using a Picloram/Triclopyr mix) are toxic to aquatic organisms and are not allowed to be used within a minimum distance from a

Box 6.6. Are frogs really good indicators?

Many people assume that the presence of frogs indicates a healthy environment. As wonderful as it is to have populations of frogs on a farm, the fact is that frogs are not always good indicators of a healthy environment. Some species readily use mud puddles and tyre tracks on the roadside and others exist in quite polluted areas. An example is the Common Eastern Froglet, which inhabits virtually any type of water body throughout south-eastern Australia.

Rather than incorrectly assuming that the occurrence of a given species of frog is an indicator species, more useful environmental information may be gained by asking questions such as: How many frog species are present? Which species are present? How many species have been lost from the area since European settlement? Are they breeding successfully? Do they show signs of environmental stress such as extra, missing or misshapen limbs?

watercourse. Therefore, it is important to choose products carefully. New, more frog-friendly 'knockdown' herbicides are being developed and several brands have already been registered for use in aquatic areas, making them suitable for use on channel banks and close to rivers, dams and creeks. Landholders should speak to their chemical supplier for more information.

A well-managed farm dam is one that has surrounding native vegetation as well as vegetation at the water margins and in the water body

Figure 6.15: Common Eastern Froglet. (Photo by Suzi Bond)

Figure 6.16: A pair of Inland Banjo Frogs. (Photo by Mason Crane)

Figure 6.17: Judy and Frank Chalker (see Box 6.7). (Photo by Mason Crane)

Box 6.7. The Chalkers' farm – a model for waterways and farm dams

Frank and Judy Chalker's farm on the South West Slopes of New South Wales provides an outstanding example of the benefits of carefully managing dams and waterways. The dams on the farm have been fenced to limit bank erosion by domestic livestock. The Chalkers also have created a set of access points to their dams and ensured that the native vegetation that has been planted around waterways is not trampled or degraded. In addition, water from the dams is siphoned (not pumped) to about 10 water troughs located at convenient watering points on their farm.

This work has resulted in a substantial increase in water quality for their livestock and greatly improved habitat quality for native wildlife on their farm – including a range of species of conservation concern like the Scarlet Robin and Flame Robin. Moreover, this approach has helped Frank and Judy Chalker's farm weather the extreme drought periods suffered by the South West Slopes between 2000 and 2010. The exciting initiatives taken on the Chalkers' farm clearly demonstrate a win–win outcome – a win for production and a win for the environment.

Summary

Waterways – streams and creeks, natural wetlands, farm dams and irrigation channels – are highly significant environments on farms.[1] When they are well managed they can provide not only high quality water for livestock but also extremely valuable habitat for a wide range of native species. They are therefore a key part of the portfolio of environmental assets on a farm. The aquatic and terrestrial environments in and around waterways are intimately interlinked and they need careful co-management. This is because the suitability of waterways for biodiversity is related both to their condition and the condition of the native vegetation around them.[10]

Just as areas of planted native vegetation are markedly different kinds of wildlife habitat to patches of remnant native vegetation (see Chapters 2 and 3), farm dams and natural waterways are also markedly different environments for frogs and birds.[7] Therefore, a well-managed farm needs to maintain the environmental integrity of both kinds of features. This requires careful whole-of-farm planning – the primary topic of the remaining chapter in this book.

References

1. Romanowski, R. 2009. *Planting Wetlands and Dams: A Practical Guide to Wetland Design, Construction and Propagation.* 2nd edn. Landlinks Press: Melbourne.
2. Martin, T.G., McIntyre, S., Catterall, C.P. and Possingham, H.P. 2006. Is landscape context important for riparian conservation? Birds in grassy woodland. *Biological Conservation* **127**: 201–214.

3. Stagoll, K., Manning, A.D., Knight, E., Fischer, J. and Lindenmayer, D.B. 2010. Using bird-habitat relationships to inform urban planning. *Landscape and Urban Planning* (in press): doi:10.1016/j.landurbplan.2010.1007.1006.
4. Sabo, J.L., Sponseller, R., Dixon, M., Gade, K., Harms, T., Heffernan, J., Jani, A., Katz, G., Soykan, C., Watts, J. and Welter, J. 2005. Riparian zones increase regional species richness by harboring different, not more, species. *Ecology* **86**: 56–62.
5. Soderquist, T.R. and Mac Nally, R. 2000. The conservation value of mesic gullies in dry forest landscapes: mammal populations in the box-ironbark ecosystem of southern Australia. *Biological Conservation* **93**: 281–291.
6. Lindenmayer, D.B. and Peakall, R. 2000. The Tumut experiment – integrating demographic and genetic studies to unravel fragmentation effects: a case study of the native bush rat. In *Genetics, Demography and Variability of Fragmented Populations.* (Eds Young, A. and Clarke, G.) pp. 173–201. Cambridge University Press: Cambridge.
7. Hazell, D., Hero, J.M., Lindenmayer, D.B. and Cunningham, R.B. 2004. A comparison of constructed and natural habitat for frog conservation in an Australian agricultural landscape. *Biological Conservation* **119**: 61–71.
8. Hazell, D., Osborne, W. and Lindenmayer, D.B. 2003. Impact of post-European stream change on frog habitat: south-eastern Australia. *Biodiversity and Conservation* **12**: 301–320.
9. Lindenmayer, D.B., Knight, E.J., Crane, M.J., Montague-Drake, R., Michael, D.R. and MacGregor, C.I. 2010. What makes an effective restoration planting for woodland birds? *Biological Conservation* **143**: 289–301.
10. Hazell, D., Cunningham, R., Lindenmayer, D., Mackey, B. and Osborne, W. 2001. Use of farm dams as frog habitat in an Australian agricultural landscape: factors affecting species richness and distribution. *Biological Conservation* **102**: 155–169.

7

What makes a good farm for biodiversity?

Introduction

This book has focused on what makes good environmental assets for farm wildlife. We have explored what makes a good remnant (Chapter 2), a good planting (Chapter 3), a good paddock (Chapter 4), a good rocky outcrop (Chapter 5), and a good waterway (Chapter 6). Individually, each of these assets has an important role to play as habitat for wildlife and, as indicated in the previous chapters in this book, any improvement in any of them will be important for wildlife on farms. These assets, however, also make an important **collective** contribution to the overall biodiversity on a farm. This final chapter focuses on what makes a good farm. We discuss the combined positive effects for biodiversity that can be generated from good management of different vegetation and other conservation assets at the farm level. We have included this chapter in the book because management at the farm level is often the most practical scale at which an individual landowner can manage a property. Yet the farm scale has only rarely received attention from biodiversity researchers who have focused either on individual patches or at the landscape scale.[1] We show that there are some important management activities at a farm scale that can make a significant difference to nature conservation, sometimes with only minor changes to existing management practices.

Of course, wildlife do not observe farm boundaries, so management practices on a neighbouring property, like fox baiting, can have a large effect on populations of native wildlife across multiple farms or landscapes and regions.[2] Given this, the second part of this chapter discusses some of the findings from our work on landscape-scale responses of biodiversity.

The importance of variation or heterogeneity in native vegetation on farms

The vegetation assets on a farm encompass many features, including patches of remnant woodland, individual paddock trees, native pastures, rocky outcrops, and creek lines. This range of assets contributes to variation in vegetation cover within and between paddocks on a farm, which is often called farm heterogeneity. Farm heterogeneity is important for biodiversity in agricultural environments around the world.[1, 3–5] We have found that many species of native animals respond strongly to the **joint** or **combined** effects of these various habitats on a farm; that is, they are significantly more likely to occur on farms where many or all of these features are present.[1] Examples include bird species like the Brown Treecreeper, Jacky Winter and Crested Shrike-tit, which are animals of conservation concern because of their substantial declines in the past few decades. Similar to the case for native birds, overall reptile species richness is greatest on farms with a high total cover of different native vegetation assets.[2] Conversely, some species which

The combined effects of different kinds of environmental assets on a farm – remnant patches, plantings, native pastures, paddock trees and rocky outcrops makes a major collective contribution to farm-level biodiversity

Figure 7.1: The Crested Shrike-tit. This charismatic bird specialises in extracting invertebrate prey from strips of eucalypt bark. It has a razor-sharp bill to tear apart bark streamers and it is often heard well before its spectacular lemon-coloured plumage is seen. (Photo by Julian Robinson)

Figure 7.2: Farms which are highly modified are more likely to support larger populations of the Eastern Brown Snake. (Photo by Suzi Bond)

are not particularly welcomed by landholders, like the Eastern Brown Snake, are less likely to occur on heterogeneous farms and appear to be more common in areas which are highly cultivated, that is, where paddock trees, native pastures and patches of remnant woodland have been removed. Highly modified farms are also more likely to support populations of the introduced House Mouse, a favoured prey item of the Eastern Brown Snake.

The characteristics of farms are important not only for birds and reptiles, but also for mammals. For example, we have found that the Common Ringtail Possum responds strongly to particular attributes of farms which we have yet to clearly elucidate.[2] We have been able to identify 'good' and 'poor' farms for possums but not the reasons underlying these effects. It seems likely, however, that farm-level practices, like excluding widespread firewood removal, benefit animals such as the Common Ringtail Possum.

Conserving biodiversity on farms – impediment or opportunity?

Some people in the Australian community consider conserving biodiversity to be an impediment to doing business – something that gets in the way of making money and getting on with agricultural production. There is another perspective

Box 7.1. My farm is extensively cleared – should I bother doing conservation work?

Throughout this book we have discussed the negative effects of vegetation clearing on biodiversity. Some landholders believe that because their farm is extensively cleared there is no point in commencing restoration work. We strongly disagree! There are opportunities to improve conservation outcomes on all farms, even highly cleared ones. For example, we have found that many species of birds and reptiles will quite rapidly colonise plantings, even on extensively cleared farms.[1, 6, 7] Interestingly, some species of birds are recruited faster to plantings on farms with low or medium amounts of remnant native vegetation than to plantings on farms with high levels of remnant native vegetation. Other studies have indicated that small-bodied native woodland birds are those most likely to be recovered in plantings on farms, including heavily cleared farms.[8]

Of course, restoration efforts can have many other benefits on farms other than helping to recover biodiversity. They can help address issues with secondary salinity, limit soil erosion, limit drought effects, and sequester large amounts of carbon (as part of tackling climate change). Active restoration programs can have spectacular results on farms. These good news stories demonstrate to farmers and the Australian public how this generation of landholders can help rewrite the nation's environmental history so that the future of agricultural zones is a more ecologically and economically sustainable one.

– that agricultural production cannot take place without biodiversity. As briefly discussed in Chapter 1, biodiversity is essential for such fundamentally crucial processes as cycling nutrients, breaking down wastes, pollinating plants, dispersing seeds and maintaining soil fertility. New government programs mean that biodiversity conservation also can help create valuable financial opportunities for farmers. These programs, like the Box-Gum Grassy Woodland Stewardship Program,[10] pay landholders to enter into long-term contracts to undertake management actions that lead to positive conservation outcomes on their farms.

There may well be other important opportunities in the future that provide financial incentives to undertake restoration programs on farms. The most notable example is carbon sequestration through planting trees. Informed revegetation programs which have carbon sequestration and biodiversity conservation as conjoint objectives offer exciting win–win opportunities for climate change mitigation and enhanced farm wildlife management.[11]

Developing a farm plan

Strong relationships between wildlife on farms and the variable portfolio or heterogeneity of vegetation assets means that the effective integration of

Box 7.2. Sand dunes at Zara – restoration by the Falkiner Group

F.S. Falkiner & Sons is the archetypal Australian pastoral company, with sprawling homesteads and gardens set on expansive country and a rich history intertwined with the development of the Australian merino industry (via the Peppin-Shaw Merino and subsequently the Poll Merino). Based in the New South Wales Riverina, near Deniliquin, it comprises nine properties spanning more than 130 000 hectares. In 2000, C, A & L Bell Commodities acquired F.S. Falkiner & Sons. They are committed to the maintenance of the properties' history and improvement of their environmental assets.

One of the most significant environmental assets on the F.S. Falkiner & Sons properties is the Zara Station sandhill. Within the Riverina bioregion, such sandhills support distinct assemblages of animals, and particularly plants. Unfortunately, the vast majority of them have been highly disturbed as a result of sand mining, rabbits, inappropriate grazing regimes by stock, and invasion by exotic weeds, particularly Spiny Burr Grass and African Boxthorn. The Zara sandhill is unique in that this 60-hectare area still closely resembles the vegetation that once covered the sandhills of the Riverina prior to European settlement. Being located near the property's homestead, the area was stocked only intermittently, and the vegetation was retained as a buffer against dust storms. The area was fenced from stock and rabbits in 1997 with support from a range of local, state and federal authorities. Pest control, particularly of rabbits and African Boxthorn has since been actively undertaken and regeneration of native tree species such as Rosewood *Alectryon oleifolis* and Sugarwood *Myoporum platycarpum* is occuring.[9]

Encouraged by these successes, F.S. Falkiner & Sons, with funding and support from the Murray Catchment Management Authority, has recently commenced restoration of other sandhills on their land. The Australian National University has established permanent monitoring points to document the change in plants and animals over time on a number of these sandhills.

agricultural production and biodiversity conservation will often be best guided by a carefully considered farm plan. An informed natural resource management agency (e.g. a Catchment Management Authority) should be able to assist landholders with creating a farm plan (see Figure 7.4 for an example). This process can be very instructive because it can help a farmer to: (1) identify places which support the different kinds of environmental assets that feature in Chapters 2–6 of this book, as well as how these areas might be best combined to create increased benefits for biodiversity; and (2) plan different options for restoration treatments in different parts of a farm (e.g. connecting previously isolated plantings or restoring degraded stream banks).

Some management activities are best coordinated using a farm plan. For example, it is clear that on farms that use irrigation water, such as those with centre-point pivot systems, clearing of paddock trees is highly likely to occur.[12] A key issue on these kinds of farms is to plan for ecologically appropriate offsetting

Box 7.3. What makes a sustainable farmer?

Life on the land can be tough and it is perhaps harder now than it has ever been. A lot has been written about sustainable agriculture, but not a lot about what makes a sustainable farmer. Being a sustainable farmer means being flexible enough to take opportunities arising from changing land-use objectives like those that may arise with carbon-based and biodiversity-based market instruments.

Other changes may include a transition for some landholders from high input–high output farming (e.g. production which relies on large amounts of fertiliser) to low input–low output farming (e.g. grazing based on extensive areas of native pasture). Other transitions may include careful consideration of the long-term future and spatial location of stands of scattered paddock trees.

These kinds of possible changes in objectives may be difficult for some people to comprehend given that it was not many decades ago that governments instructed landholders to clear vast areas of land and add large quantities of fertiliser for paddock improvement (e.g. in south-western Western Australia in the 1970s and 1980s). Nevertheless, it is critically important for modern landholders to be aware of these kinds of changes and take advantage of opportunities they create.

There can be major psychological benefits associated with good environmental management which contribute to the sustainability of farming life. Many landowners note that the greenery of tree plantings, listening to bird calls, or watching the recovery of remnant vegetation along a once degraded creek, are the type of things that inspire positive thinking when times are tough.

so that where the loss of paddock trees is unavoidable on one part of a farm, this loss is more than compensated for by marked conservation improvement elsewhere on the same farm.[13] This kind of whole-of-farm planning and offsetting, however, needs very careful thinking to avoid perverse outcomes, such as improvements in some woodland patches but significant environmental degradation across the rest of a farm.[14]

It is not possible to manage for all things on every hectare of a farm. Different priorities will be appropriate in different parts of a farm. A farm plan is useful to help determine which actions are good ones in which places

The use of prescribed fire is another kind of management activity that is best guided by a whole-of-farm plan. This is because: (1) an appropriate fire regime is dependent on the objectives of burning (e.g. controlling exotic plants versus promoting the germination of particular native species); and (2) the responses of biodiversity to fire are extremely variable and complex – what benefits some species will be highly detrimental to others. As a general rule of thumb, vegetation ecologists recommend that the application of prescribed fire is most appropiate in woodlands that have been historically burned.[15] In addition, a mosaic of burned and unburned

Box 7.4. An audit of environmental assets on a farm

It is hard to plan unless you know what you've got. Some farmers may be unaware of the environmental assets on a farm, how important they are, and how to best manage them. A map of the location and extent of key assets like rocky outcrops, patches of woodland remnants, areas of native pasture and scattered paddock trees can assist greatly with the process of farm planning and then developing priorities for management action. Regional bodies like Catchment Management Authorities can help landholders create a map of the environmental assets on their farm that will, in turn, form the basis for a good farm plan. We believe there is a need for governments and agencies involved in natural resource management not to penalise landholders for good management, for example, by increasing regulatory obligations should a property be improved for a threatened species and populations of that species increase as a result.

areas will often be best for promoting post-fire ecological recovery of many species. It is therefore best to plan in advance where fire will be excluded (e.g. wetlands and riparian areas), as well as which parts of paddocks and remnants will be burned – after giving careful consideration to factors like the time elapsed since those areas were last burned and the severity of past fires.

The application of fertiliser is yet another activity on farms that might be best managed at the farm level. In various parts of this book, we have discussed how large amounts of fertiliser have negative effects on the diversity of native plants, increase exotic plant species, contribute to dieback in paddock trees, and contribute to the demise of the integrity of the soil layer. There also appear to be negative effects of fertiliser on bird diversity. Given these effects, we suggest there should be parts of farms where there is no application of artificial fertiliser.

Beyond the farm – the landscape scale

Some important ecological processes in agricultural areas are landscape-scale or regional-scale processes. (Here, we refer to landscape-scale as the scale larger than the single farm-level.) For example, recent work is suggesting that large-scale tree clearing in the Murray–Darling Basin has contributed to the depth and extent of recent drought events.[16] Notably, during the early 1800s, many landholders in districts like the Monaro on the New South Wales Southern Tablelands believed that trees brought rain and so clearing was not extensive.[17] Unfortunately, this did not deter subsequent widespread land clearing.

Many species of plants and animals respond to the characteristics of the environment at several different spatial scales. As an example, the Brown Treecreeper is often found in woodland patches with large amounts of fallen

Figure 7.3: (a) Faye and Bill Belling, (b) and (c) the Bellings' farm near Tarcutta on the South West Slopes of New South Wales. In the 1980s, this farm suffered many problems with rising water tables and extensive soil erosion. An active restoration program was instigated and this has had major positive benefits for arresting secondary salinity and improving farm productivity. The Bellings' farm is also a significant property for a range of declining woodland birds, including the Hooded Robin, Crested Shrike-tit and the Black-chinned Honeyeater. (Photos by Sachiko Okada)

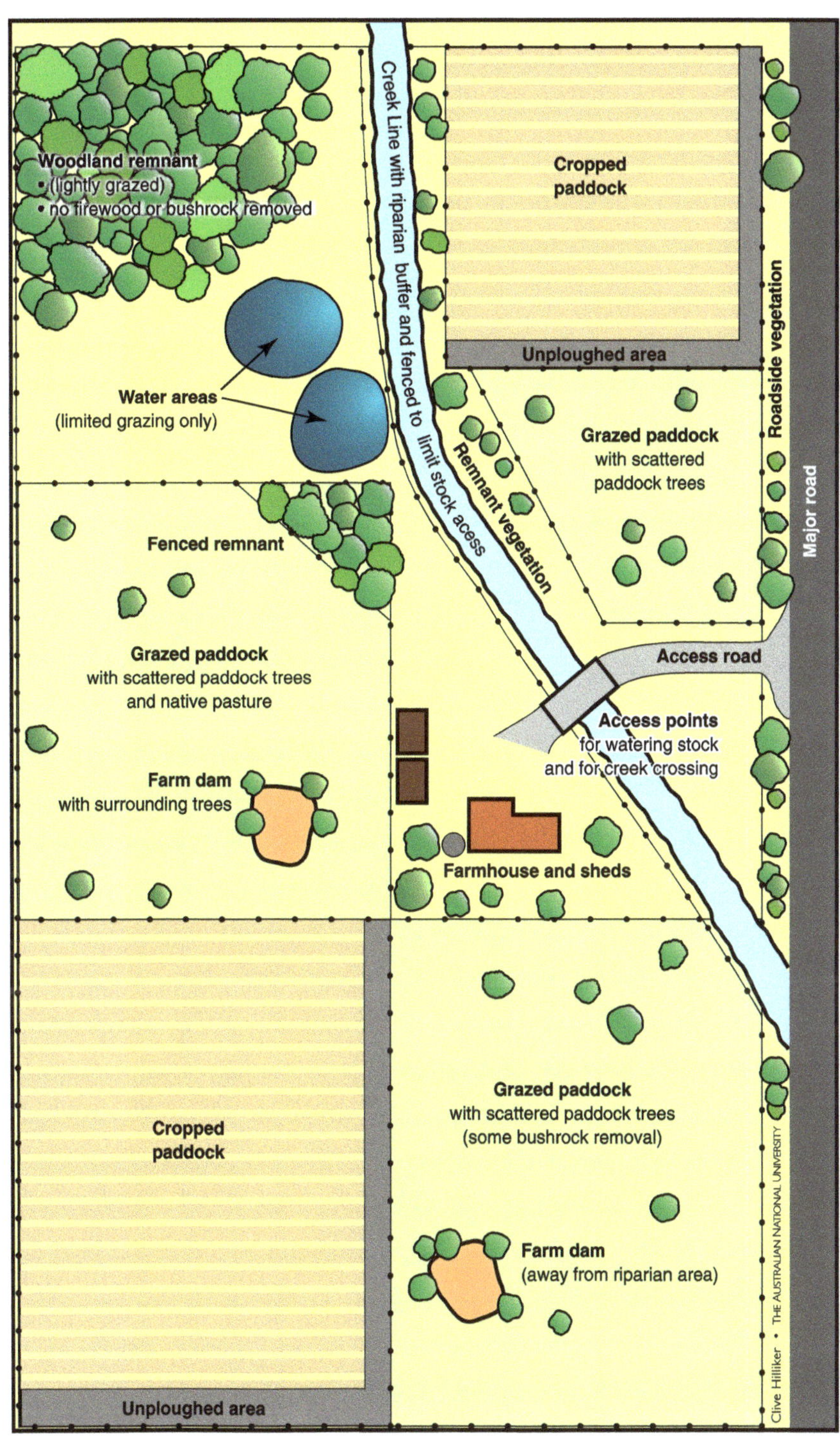

Figure 7.4: A hypothetical farm plan. (Drawing by Clive Hilliker)

timber. It also requires large patches of woodland. In addition, patches occupied by the species are those which are surrounded by many other large patches of woodland.[18] Therefore, attributes of sites, patches, farms and landscapes (i.e. multiple farms) combine to influence the occurrence of this species.

The Brown Treecreeper is far from an isolated case and illustrates the influence on animal distribution of factors at multiple scales. We have found many other species ranging from reptiles to possums and an array of other bird species (e.g. the White-plumed Honeyeater and the Superb Parrot) respond to features of the environment at a range of scales. An understanding of scale effects is important for conservation because it highlights the appropriate scale or scales at which management actions are likely to be most effective. A good example is baiting for feral predators like the Red Fox. Work on movement patterns and predation effects indicate that landscape-scale baiting programs are required to control this serious environmental pest. Therefore, the most effective bating programs will be those coordinated across several adjoining farms. The value of doing this appears to be underscored by studies of species such as the Common Brushtail Possum, which are often preyed upon by the Red Fox. The Common Brushtail Possum is more likely to occur in landscapes where there is an ongoing tradition of coordinating fox baiting across many farms.[2]

Figure 7.5: The Brown Treecreeper is a bird of conservation concern which responds to characteristics at several scales – sites, patches, farms and landscapes. (Photo by Julian Robinson)

Finally, the total amount of native vegetation cover in a landscape or region can have an important influence on the occurrence of some species of animals in temperate woodland environments. Effective conservation requires balancing different kinds of actions across different scales in a complementary manner. Therefore, actions like establishing plantings, fencing individual remnants, and promoting natural regeneration of trees within paddocks need to be complemented by strategies at larger landscape scales that are designed to increase the total extent of native vegetation on farms[1] and in landscapes.[19]

Scientific research has documented relationships between the bird communities and the vegetation cover of agricultural landscapes.[8, 20] Landscapes that include both remnants and plantings (with shrubs and understorey plants) support more diverse assemblages of birds, particularly small-bodied woodland birds. In landscapes with a more simplified cover of native vegetation, these birds decline and larger-bodied species (such as the Noisy Miner, Galah and Australian Magpie) tend to dominate. These findings are consistent with the findings of the work on the importance for native birds of understorey vegetation in remnants[18] (see Chapter 2) and in plantings[21] (Chapter 3).

Some kinds of management are best co-ordinated across sets of farms. Fox baiting is a good example

Figure 7.6: Fox baiting is a key farm management practice that is most effective if coordinated across many farms. (Photo by David Lindenmayer)

Figure 7.7: Buff-rumped Thornbill. An example of a smaller-bodied species of woodland bird attracted to plantings. (Photo by Suzi Bond)

Travelling stock reserves as critical parts of agricultural landscapes for biodiversity conservation

Some parts of agricultural landscapes are particularly important for biodiversity. In New South Wales and Queensland, these include travelling stock reserves. The equivalent of travelling stock reserves in Victoria include Crown reserves and roadside reserves. The travelling stock reserve network was established more than 150 years ago to facilitate the movement of domestic livestock between properties and markets. Travelling stock reserves have many important cultural and heritage values, for both indigenous and European people.[22–24] Because of their periodic use and the fact that they generally escaped vegetation clearing and prolonged set-stocking of livestock,[25–27] travelling stock reserves are often also valuable for biodiversity conservation. Based on studies of travelling stock reserves in the Murray and Murrumbidgee regions of southern New South Wales, we found that in comparison with temperate woodland remnants in other land tenures, travelling stock reserves tend to support more species of birds, more species of declining birds, and a greater abundance of arboreal marsupials. Travelling stock reserves also are generally more likely to be occupied by particular species of birds of conservation concern. These include the Red-capped Robin, Grey-crowned Babbler, Rufous Whistler and the Brown Treecreeper.

Although our work has shown that travelling stock reserves are an important resource for biodiversity conservation, this does not mean that temperate

Figure 7.8: The Grey-crowned Babbler is more likely to occupy travelling stock reserves than woodland patches in other land tenures in South West Slopes region. (Photo by Julian Robinson)

woodlands under different tenures are also not valuable. Travelling stock reserves should be viewed as a complementary resource rather than a potential replacement for existing private woodland remnants and other reserves. Travelling stock reserves should be considered as one of several kinds of temperate woodland areas which collectively contribute to a portfolio of vegetation assets in formerly woodland-dominated, but now extensively cleared, landscapes. The paucity of large ecological reserves in the Murray Riverina and South West Slopes bioregions[28] means that the collective contribution of land under different tenures is likely to be vitally important for biodiversity.

As yet, there are no appropriate data to quantify the collective value of interconnected sets of travelling stock reserves as networks to promote animal and plant movement throughout landscapes and hence their regional and/or national contribution to biodiversity conservation. However, it is likely that travelling stock reserves will have a valuable network role for some species.[29]

Summary

A good farm for biodiversity needs to consider the key habitats that a farm can support – remnant native vegetation, plantings, paddocks, rocky outcrops, creeks and farm dams. Each of these assets is important in its own right but they are also collectively important for biodiversity on a farm. The variation and diversity of

Figure 7.9: (a) Britta's Reserve, a travelling stock reserve that provides important habitat for the highly threatened Bush Stone Curlew. (Photo by David Lindenmayer). (b) Bush Stone Curlew. (Photo by Esther Beaton)

assets is also important as it contributes to farm-level and landscape-level heterogeneity which has been found to be important for maintaining biodiversity in farming environments around the world.[1, 3, 8]

Conclusion

Australia's biodiversity is a critical part of our nation's heritage. However, a great many unique species of plants and animals are now threatened or endangered. We have shown throughout this book that good farm management - whether it be in woodland remnants, plantings, paddocks, rocky outcrops or waterways - can make an enormous positive contribution to biodiversity protection and conservation. There have been many success stories – the recovery of bird populations like the Grey-crowned Babbler as a result of planting efforts is one of many very positive examples. We hope that the new research and management insights discussed in this book will not only further assist landholders already doing important conservation work on their farms, but also encourage others to join in and contribute to this very important work.

References

1. Cunningham, R.B., Lindenmayer, D.B., Crane, M., Michael, D.R., MacGregor, C., Montague-Drake, R. and Fischer, J. 2008. The combined effects of remnant vegetation and tree planting on farmland birds. *Conservation Biology* **22**: 742–752.
2. Cunningham, R.B., Lindenmayer, D.B., Crane, M., Michael, D. and MacGregor, C. 2007. Reptile and arboreal marsupial response to replanted vegetation in agricultural landscapes. *Ecological Applications* **17**: 609–619.
3. Benton, T.G., Vickery, J.A. and Wilson, J.D. 2003. Farmland biodiversity: is habitat heterogeneity the key? *Trends in Ecology and Evolution* **18**: 182–188.
4. Laiolo, P. 2004. Spatial and seasonal patterns of bird communities in Italian agroecosystems. *Conservation Biology* **19**: 1547–1556.
5. Sekercioglu, C.H., Loarie, S.C., Brenes, F.O., Ehrlich, P.R. and Daily, G.C. 2007. Persistence of forest birds in the Costa Rican agricultural countryside. *Conservation Biology* **21**: 482–494.
6. Munro, N., Lindenmayer, D.B. and Fischer, J. 2007. Faunal response to revegetation in agricultural areas of Australia: a review. *Ecological Management and Restoration* **8**: 199–207.
7. Munro, N.T., Fischer, J., Wood, J. and Lindenmayer, D.B. 2009. Revegetation in agricultural areas: the development of structural complexity and floristic diversity. *Ecological Applications* **19**: 1197–1210.

8. Fischer, J., Lindenmayer, D.B. and Montague-Drake, R. 2008. The role of landscape texture in conservation biogeography: a case study on birds in south-eastern Australia. *Diversity and Distributions* **14**: 38–46.
9. Stafford, M. and Eldridge, D.J. 2000. Vegetation, soils and management of 'Zara': a sandhill remnant on the Riverine Plain. *Cunninghamia* **6**: 717–746.
10. Department of the Environment, Water, Heritage and the Arts. 2009. 'Environmental Stewardship Program. Box-Gum Grassy Woodland Project'. Department of the Environment, Water, Heritage and the Arts: Canberra.
11. Bekessy, S.A. and Wintle, B.A. 2008. Using carbon investment to grow the biodiversity bank. *Conservation Biology* **22**: 510–513.
12. Maron, M. and Fitzsimons, J.A. 2007. Agricultural intensification and loss of matrix habitat over 23 years in the West Wimmera, south-eastern Australia. *Biological Conservation* **135**: 587–593.
13. Gibbons, P. and Lindenmayer, D.B. 2007. Offsets for land clearing: no net loss or the tail wagging the dog? *Environmental Management and Restoration* **8**: 26–31.
14. Spooner, P.G. and Briggs, S.V. 2008. Woodlands on farms in southern New South Wales: a longer-term assessment of vegetation changes after fencing. *Ecological Management and Restoration* **9**: 33–41.
15. Prober, S.M., Thiele, K.R. and Lunt, I.D. 2007. Fire frequency regulates tussock grass composition, structure and resilience in endangered temperate woodlands. *Austral Ecology* **32**: 808–824.
16. McAlpine, C.A., Syktus, J., Deo, R.C., Lawrence, P.J., McGowan, H.A., Watterson, I.G. and Phinn, S.R. 2007. Modeling the impact of historial land cover change on Australia's regional climate. *Geophysical Research Letters* **34**: L22711.1-L22711.6.
17. Hancock, K. 1972. *Discovering Monaro: A Study of Man's Impact on His Environment.* Cambridge University Press: Cambridge.
18. Montague-Drake, R.M., Lindenmayer, D.B. and Cunningham, R.B. 2009. Factors affecting site occupancy by woodland bird species of conservation concern. *Biological Conservation* **142**: 2896–2903.
19. Radford, J.Q., Bennett, A.F. and Cheers, G.J. 2005. Landscape-level thresholds of habitat cover for woodland-dependent birds. *Biological Conservation* **124**: 317–337.
20. Fischer, J., Lindenmayer, D.B., Blomberg, S., Montague-Drake, R. and Felton, A. 2007. Functional richness and relative resilence of bird communities in regions with different land use intensities. *Ecosystems* **10**: 964–974.
21. Lindenmayer, D.B., Knight, E.J., Crane, M.J., Montague-Drake, R., Michael, D.R. and MacGregor, C.I. 2010. What makes an effective restoration planting for woodland birds? *Biological Conservation* **143**: 289–301.

22. Spooner, P.G. 2005. On squatters, settlers and early surveyors: historical development of country road reserves in southern New South Wales. *Australian Geographer* **36**: 55–73.
23. Reynolds, H. 1990. *With the White People*. Penguin: Melbourne.
24. Byrne, D. and Harrsion, R. 2004. 'Aboriginal and shared heritage in TSRs'. NSW Department of Environment and Climate Change: Hurstville, Sydney.
25. Eddy, D. 2002. *Managing Native Grasslands. A Guide to Management for Conservation, Production and Landscape Protection*. WWF Australia: Sydney.
26. Davidson, I., Scammell, A., O'Shannassy, P., Mullins, M. and Learmonth, S. 2005. Travelling stock reserves: refuges for stock and biodiversity? *Ecological Management and Restoration* **6**: 5–15.
27. Spooner, P.G. and Lunt, I. 2004. The influence of land-use history on roadside conservation values in an Australian agricultural landscape. *Australian Journal of Botany* **52**: 445–458.
28. Pressey, R.L., Hager, T.C., Ryan, K.M., Wall, S., Ferrier, S. and Creaser, P.M. 2000. Using abiotic data for conservation assessments over extensive regions: quantitative methods applied across New South Wales, Australia. *Biological Conservation* **96**: 55–82.
29. Possingham, H.P. and Nix, H.A. 2008. The Long Paddock Scientists' Statement. http://www.wilderness.org.au/files/long-paddock-scientists-statement.pdf.

Index

www.ingramcontent.com/pod-product-compliance
Lightning Source LLC
LaVergne TN
LVHW060631110826
845147LV00014B/888

* 9 7 8 0 6 4 3 1 0 0 3 1 2 *